Introduction to Modeling and Simulation with MATLAB® and Python

Chapman & Hall/CRC
Computational Science Series

SERIES EDITOR

Horst Simon
Deputy Director
Lawrence Berkeley National Laboratory
Berkeley, California, U.S.A.

PUBLISHED TITLES

COMBINATORIAL SCIENTIFIC COMPUTING
Edited by Uwe Naumann and Olaf Schenk

CONTEMPORARY HIGH PERFORMANCE COMPUTING: FROM PETASCALE
TOWARD EXASCALE
Edited by Jeffrey S. Vetter

CONTEMPORARY HIGH PERFORMANCE COMPUTING: FROM PETASCALE
TOWARD EXASCALE, VOLUME TWO
Edited by Jeffrey S. Vetter

DATA-INTENSIVE SCIENCE
Edited by Terence Critchlow and Kerstin Kleese van Dam

ELEMENTS OF PARALLEL COMPUTING
Eric Aubanel

THE END OF ERROR: UNUM COMPUTING
John L. Gustafson

EXASCALE SCIENTIFIC APPLICATIONS: SCALABILITY AND
PERFORMANCE PORTABILITY
Edited by Tjerk P. Straatsma, Timothy J. Williams, and Katerina Antypas

FROM ACTION SYSTEMS TO DISTRIBUTED SYSTEMS: THE REFINEMENT APPROACH
Edited by Luigia Petre and Emil Sekerinski

FUNDAMENTALS OF MULTICORE SOFTWARE DEVELOPMENT
Edited by Victor Pankratius, Ali-Reza Adl-Tabatabai, and Walter Tichy

FUNDAMENTALS OF PARALLEL MULTICORE ARCHITECTURE
Yan Solihin

THE GREEN COMPUTING BOOK: TACKLING ENERGY EFFICIENCY AT LARGE SCALE
Edited by Wu-chun Feng

GRID COMPUTING: TECHNIQUES AND APPLICATIONS
Barry Wilkinson

HIGH PERFORMANCE COMPUTING: PROGRAMMING AND APPLICATIONS
John Levesque with Gene Wagenbreth

PUBLISHED TITLES CONTINUED

PUBLISHED TITLES CONTINUED

Introduction to Modeling and Simulation with MATLAB® and Python

Steven I. Gordon

Brian Guilfoos

CRC Press

Taylor & Francis Group

Boca Raton London New York

CRC Press is an imprint of the
Taylor & Francis Group, an **informa** business

A CHAPMAN & HALL BOOK

MATLAB® and Simulink® are trademarks of the MathWorks, Inc. and are used with permission. The MathWorks does not warrant the accuracy of the text or exercises in this book. This book's use or discussion of MATLAB® and Simulink® software or related products does not constitute endorsement or sponsorship by the MathWorks of a particular pedagogical approach or particular use of the MATLAB® and Simulink® software.

CRC Press
Taylor & Francis Group
6000 Broken Sound Parkway NW, Suite 300
Boca Raton, FL 33487-2742

© 2017 by Taylor & Francis Group, LLC
CRC Press is an imprint of Taylor & Francis Group, an Informa business

No claim to original U.S. Government works

Printed on acid-free paper

International Standard Book Number-13: 978-1-4987-7387-4 (Hardback)

Visit the Taylor & Francis Web site at
http://www.taylorandfrancis.com

and the CRC Press Web site at
http://www.crcpress.com

Contents

Preface

ODELING AND SIMULATION USING computation or computational
science has become an essential part of the research and development process in the physical, biological, and social sciences and engineering. It allows the exploration of physical and biological systems at the micro- and molecular level that increase our understanding of their function and the discovery of new materials and new drugs. It allows us to understand the interactions of components in complex systems from those we engineer and build to our ecosystems and climate. In recent years, computational science has produced enormous advances in almost all fields of scientific and technological inquiry, including DNA sequencing, behavioral modeling, global climatic predictions, drug design, financial systems, and medical visualization. At the same time, it has become critical in the design, testing, and manufacturing of new products and services, saving millions of dollars in development costs and getting new products to market more rapidly.

Scientists, social scientists, and engineers must have an understanding of both modeling and computer programming principles so that they appropriately apply those techniques in their practice. Several sets of knowledge and skills are required to achieve that understanding. How do we translate the relationships within a system being modeled into a set of mathematical functions that accurately portray the behavior of that system? How are the mathematics translated into computer code that correctly simulates those relationships? What is the nature of errors introduced by simplifying the depiction of the system, introduced by the computer algorithm used to solve the equations, and limited by our knowledge of the system behavior? How accurate is the model? How do we know the model is logically correct and follows from the physical and mathematical laws used to create it (verification)? How do we demonstrate that the model correctly predicts the phenomena modeled (validation)? These are the underlying questions that are the focus of this book.

The book is intended for students and professionals in science, social science, and engineering who wish to learn the principles of computer modeling as well as basic programming skills. For many students in these fields, with the exception of computer science students and some engineering students, enrollment in an introductory programming course may be impractical or difficult. At many institutions, these courses are focused primarily on computer science majors and use a programming language such as Java that is not readily applicable to science and engineering problems. We have found that teaching programming as a just-in-time tool used to solve real problems more deeply engages those students to master the programming concepts. Combining that effort with learning the principles of modeling and simulation provides the link between programming and problem solving while also fitting more readily into a crowded curriculum.

For students from all fields, learning the basic principles of modeling and simulation prepares them for understanding and using computer modeling techniques that are being applied to a myriad of problems. The knowledge of the modeling process should provide the basis for understanding and evaluating models in their own subject domain. The book content focuses on meeting a set of basic modeling and simulation competencies that were developed as part of several National Science Foundation grants (see http://hpcuniversity.org/educators/undergradCompetencies/). Even though computer science students are much more expert programmers, they are not often given the opportunity to see how those skills are being applied to solve complex science and engineering problems, and may also not be aware of the libraries used by scientists to create those models.

We have chosen to use MATLAB® and Python for several reasons. First, both offer interfaces that the intended audience should find intuitive. Both interfaces provide instant feedback on syntax errors and extensive help documents and tutorials that are important for novice programmers. Although MATLAB is a commercially licensed program, whereas Python is open source, many campuses currently have a site license for MATLAB. Students can also purchase the student version of MATLAB relatively cheaply.

Perhaps most importantly, both programs are extensively used by the science and engineering community for model development and testing. Even though neither program scales as efficiently as C, Fortran, or other languages for large-scale modeling on current parallel comput-ing architectures, they do offer a stepping stone to those environments. Both have extensive toolkits and scientific and mathematical libraries that can be invoked to reduce the amount of coding required to undertake many modeling projects. Although we use these programming environ-ments to teach rudimentary programming techniques without applying a large number of these tools, they are available to students for develop-ing capstone projects or for use in more advanced courses later in their curriculum.

ORGANIZATION OF THE BOOK

The book interleaves chapters on modeling concepts and related exercises with programming concepts and exercises. We start out with an introduc-tion to modeling and its importance to current practices in the sciences and engineering. We then introduce each of the programming environ-ments and the syntax used to represent variables and compute math-ematical equations and functions. As students gain more programming expertise, we go back to modeling concepts, providing starting code for a variety of exercises where students add additional code to solve the prob-lem and provide an analysis of the outcomes. In this way, we build both modeling and programming expertise with a "just-in-time" approach so that by the end of the book, students can take on relatively simple model-ing example on their own.

Each chapter is supplemented with references to additional read-ing, tutorials, and exercises that guide students to additional help and allow them to practice both their programming and analytical modeling skills. The companion website at http://www.intromodeling.com provides updates to instructions when there are substantial changes in software versions as well as electronic copies of exercises and the related code. Solutions to the computer exercises are available to instructors on the publisher's website.

Each of the programming-related chapters is divided into two parts—one for MATLAB and one for Python. We assume that most instructors will choose one or the other so that students can focus only on the lan-guage associated with their course. In these chapters, we also refer to addi-tional online tutorials that students can use if they are having difficulty with any of the topics.

The book culminates with a set of final project exercise suggestions that incorporate both the modeling and the programming skills provided in the rest of the volume. These projects could be undertaken by individuals or small groups of students. They generally involve research into a particular modeling problem with suggested background reading from the literature. Each exercise has a set of starting code providing a very simplistic view of the system and suggestions for extending the model by adding additional components to relax some of the assumptions. Students then complete the program code and use the model to answer a number of questions about the system, complete model verification and validation where possible, and present a report in written and oral form.

The website also offers a space where people can suggest additional projects they are willing to share as well as comments on the existing projects and exercises throughout the book. We hope that the combination of materials contributes to the success of those interested in gaining modeling and simulation expertise.

MATLAB® is a registered trademark of The MathWorks, Inc. For product information, please contact:

The MathWorks, Inc.
3 Apple Hill Drive
Natick, MA 01760-2098 USA
Tel: 508 647 7000
Fax: 508-647-7001
E-mail: info@mathworks.com
Web: www.mathworks.com

Authors

Steven I. Gordon is a professor emeritus of the City and Regional Planning and Environmental Science Programs at the Ohio State University, Columbus, Ohio. He earned a bachelor's degree from the University at Buffalo, Buffalo, New York, in 1966 and a PhD degree from Columbia University, New York, in 1977. He also serves as the senior education lead at the Ohio Supercomputer Center. In that and other roles at OSC, he has focused primarily on the integration of computational science into the curricula at higher education institutions in Ohio and throughout the United States. He has worked with multiple institutions through a variety of grants from the National Science Foundation, including the Extreme Science and Engineering Discovery Environment (XSEDE) and Blue Waters project.

Dr. Gordon is also one of the founders and first chair of the Association of Computing Machinery (ACM) Special Interest Group High Performance Computing (SIGHPC) Education Chapter and serves as a representative of the SIGHPC on the ACM Education Council. He has published extensively on topics related to environmental planning and the applications of modeling and simulation in education and research.

Brian Guilfoos serves as the High Performance Computing (HPC) Client Services manager for the Ohio Supercomputer Center (OSC), Columbus, Ohio. Guilfoos leads the HPC Client Services Group, which provides training and user support to facilitate the use of computational science by the center's user communities. He earned a master's degree in public policy and administration in 2014 and a bachelor's degree in electrical engineering in 2000, both from the Ohio State University, Columbus, Ohio. He also works directly with OSC clients to help convert computer codes and develop batch scripting, compiling, and code development so that these researchers can efficiently use the center's supercomputers and licensed software.

Guilfoos developed and delivered training in MATLAB® as a part of the U.S. Department of Defense High Performance Computing Modernization Program support.

Prior to joining OSC, he was contracted by the Air Force Research Laboratory (AFRL) to focus on software development in support of unmanned aerial vehicle interface research. He was a key technical member of a team that was awarded the 2004 Scientific and Technological Achievement Award by the AFRL's Human Effectiveness Directorate.

CHAPTER 1

Introduction to Computational Modeling

1.1 THE IMPORTANCE OF COMPUTATIONAL SCIENCE

Advances in science and engineering have come traditionally from the application of the scientific method using theory and experimentation to pose and test our ideas about the nature of our world from multiple perspectives. Through experimentation and observation, scientists develop theories that are then tested with additional experimentation. The cause and effect relationships associated with those discoveries can then be represented by mathematical expressions that approximate the behavior of the system being studied.

With the rapid development of computers, scientists and engineers translated those mathematical expressions into computer codes that allowed them to imitate the operation of the system over time. This process is called simulation. Early computers did not have the capability of solving many of the complex system simulations of interest to scientists and engineers. This led to the development of supercomputers, computers with higher level capacity for computation compared to the general-purpose computers of the time. In 1982, a panel of scientists provided a report to the U.S. Department of Defense and the National Science Foundation urging the government to aid in the development of supercomputers (Lax, 1982). They indicated that "the primacy of the U.S. in science, engineering, and computing technology could be threatened relative to that of other countries with national efforts in supercomputer access and development." They recommended both investments in research and development and in the training of personnel in science and engineering computing.

The capability of the computer chips in your cell phone today far exceeds that of the supercomputers of the 1980s. The Cray-1 supercomputer released in 1975 had a raw computing power of 80 million floating-point operations per second (FLOPS). The iPhone 5s has a graphics processor capable of 76.8 Gigaflops, nearly one thousand times more powerful (Nick, 2014). With that growth in capability, there has been a dramatic expansion in the use of simulation for engineering design and research in science, engineering, social science, and the humanities. Over the years, that has led to many efforts to integrate computational science into the curriculum, to calls for development of a workforce prepared to apply computing to both academic and commercial pursuits, and to investments in the computer and networking infrastructure required to meet the demands of those applications. For example, in 2001 the Society for Industrial and Applied Mathematics (SIAM) provided a review of the graduate education programs in science and engineering (SIAM, 2001). They defined computational science and engineering as a multidisciplinary field requiring expertise in computer science, applied mathematics, and a subject field of science and engineering. They provided examples of emerging research, an outline of a curriculum, and curriculum examples from both North America and Europe.

Yasar and Landau (2001) provided a similar overview of the interdisciplinary nature of the field. They also describe the possible scope of programs at the both the undergraduate and graduate levels and provide a survey of existing programs and their content. More recently, Gordon et al. (2008) described the creation of a competency-based undergraduate minor program in computational science that was put into place at several institutions in Ohio. The competencies were developed by an interdisciplinary group of faculty and reviewed by an industry advisory committee from the perspective of the skills that prospective employers are looking for in students entering the job market. The competencies have guided the creation of several other undergraduate programs. They have also been updated and augmented with graduate-level computational science competencies and competencies for data-driven science. The most recent version of those competencies can be found on the HPC University website (HPC University, 2016).

More recently, there have been a number of national studies and panels emphasizing the need for the infrastructure and workforce required to undertake large-scale modeling and simulation (Council on Competitiveness, 2004; Joseph et al., 2004; Reed, 2005; SBES, 2006). This book provides an introduction to computational science relevant to students across the spectrum of science and engineering. In this chapter, we begin with a brief review of the history or computational modeling and its contributions to the advancement of science. We then provide an overview of the modeling process and the terminology associated with modeling and simulation.

As we progress through the book, we guide students through basic programming principles using two of the widely used simulation environments—MATLAB® and Python. Each chapter introduces either a new set of programming principles or applies them to the solution of one class of models. Each chapter is accompanied by exercises that help to build both basic modeling and programming skills that will provide a background for more advanced modeling courses.

1.2 HOW MODELING HAS CONTRIBUTED TO ADVANCES IN SCIENCE AND ENGINEERING

There are a myriad of examples documenting how modeling and simulation has contributed to research and to the design and manufacture of new products. Here, we trace the history of computation and modeling to illustrate how the combination of advances in computing hardware, software, and scientific knowledge has led to the integration of computational modeling techniques throughout the sciences and engineering. We then provide a few, more recent examples of advances to further illustrate the state-of-the-art. One exercise at the end of the chapter provides an opportunity for students to examine additional examples and share them with their classmates.

The first electronic programmable computer was the ENIAC built for the army toward the end of World War II as a way to quickly calculate artillery trajectories. Herman Goldstine (1990), the project leader, and two professors from the University of Pennsylvania, J. Presper Eckert, and John Mauchly sold the idea to the army in 1942 (McCartney, 1999). As the machine was being built and tested, a large team of engineers and mathematicians was assembled to learn how to use it. That included six women mathematicians who were recruited from colleges across the country. As the machine was completed in 1945, the war was near an end.

ENIAC was used extensively by the mathematician John von Neumann not only to undertake its original purposes for the army but also to create the first weather model in 1950. That machine was capable of 400 floating-point operations per second and needed 24 hours to calculate the simple daily weather model for North America. To provide a contrast to the power of current processors, Peter and Owen Lynch (2008) created a version of the model that ran on a Nokia 6300 mobile phone in less than one second!

It is impossible to document all of the changes in computational power and its relationship to the advancements in science that have occurred since this first computer. Tables 1.1 and 1.2 show a timeline

TABLE 1.1 Timeline of Advances in Computer Power and Scientific Modeling (Part 1)

Example Hardware	Max. Speed	Date	Weather and Climate Modeling
ENIAC	400 Flops	1945	
		1950	First automatic weather forecasts
UNIVAC		1951	
IBM 704	12 KFLOP	1956	
		1959	Ed Lorenz discovers the chaotic behavior of meteorological processes
IBM7030 Stretch; UNIVAC LARC	500-500 KFLOP	~1960	
		1965	Global climate modeling underway
CDC6600	1 Megaflop	1966	
CDC7600	10 MFLOP	1975	
CRAY1	100 MFLOP	1976	
CRAY-X-MP	400 MFLOP		
		1979	Jule Charney report to NAS
CRAY Y-MP	2.67 GFLOP		
		1988	Intergovernmental Panel on Climate Change
		1992	UNFCCC in Rio
IBM SP2	10 Gigaflop	1994	
ASCII Red	2.15 TFLOP	1995	Coupled Model Intercomparison Project (CMIP)
		2005	Earth system models
Blue Waters	13.34 PFLOP	2014	

Sources: Bell, G., Supercomputers: The amazing race (a history of supercomputing, 1960–2020), 2015, http://research.microsoft.com/en-us/um/people/gbell/MSR-TR-2015-2_Supercomputers-The_Amazing_Race_Bell.pdf (accessed December 15, 2016).
Bell, T., Supercomputer timeline, 2016, https://mason.gmu.edu/~tbell5/page2.html (accessed December 15, 2016).
Esterbrook, S., Timeline of climate modeling, 2015, https://prezi.com/pakaaiek3nol/timeline-of-climate-modeling/ (accessed December 15, 2016).

TABLE 1.2   Timeline of Advances in Computer Power and Scientific Modeling (Part 2)

Date	Theoretical Chemistry	Aeronautics and Structures	Software and Algorithms
1950	Electronic wave functions		
1951	Molecular orbital theory (Roothan)		
1953	One of the first molecular simulations (Metropolis et al.)		
1954			Vector processing directives
1956	First calculation of multiple electronic states of a molecule on EDSAC (Boys)		
1957			FORTRAN created
1965	Creation of ab initio molecular modeling (People)		
1966		2D Navier-Stokes simulations; FLO22; transonic flow over a swept wing	
1969			UNIX created
1970		2D Inviscid Flow Models; design of regional jet	
1971		Nastran (NASA Structural Analysis)	
1972			C programming language created
1973			Matrix computations and errors (Wilkinson)
1975		3D Inviscid Flow Models; complete airplane solution	
1976	First calculation of a chemical reaction (Warshel)	DYNA3D which became LS-DYNA (mid-70s)	
1977	First molecular dynamics of proteins (Karplus) First calculation of a reaction transition state (Chandler)	Boeing design of 737-500	

(*Continued*)

TABLE 1.2 (*Continued*)   Timeline of Advances in Computer Power and Scientific Modeling (Part 2)

Date	Theoretical Chemistry	Aeronautics and Structures	Software and Algorithms
1979			Basic Linear Algebra Subprograms (BLAS) library launched
1980s	*Journal of Computational Chemistry* first published	800,000 mesh cells around a wing, FLO107	
1984			MATLAB created
1985		Design of Boeing 767,777	GNU project launched (free Software foundation)
1991			Linux launched
1993			Message passing interface (MPI) specification
1994			Python created
1995	First successful computer-based drug design (Kubinyi)		
1997			Open multiprocessing (OpenMP) specification
2000		Discontinuous finite element methods; turbulent flow; design of airbus	
2007			CUDA launched
2014			Open accelerator (OpenACC) specification

Sources: Bartlett, B.N., The contributions of J.H. Wilkinson to numerical analysis. In S.G. Nash, (Ed.), *A History of Scientific Computing*, ACM Press, New York, pp. 17–30, 1990.

Computer History Museum, Timeline of computer history, software and languages, 2017, http://www.computerhistory.org/timeline/software-languages/ (accessed January 2, 2017).

Dorzolamide, 2016, https://en.wikipedia.org/wiki/Dorzolamide (accessed December 15, 2016).

Jameson, A., Computational fluid dynamics, past, present, and future, 2016, http://aero-comlab.stanford.edu/Papers/NASA_Presentation_20121030.pdf (accessed December 15, 2016).

Prat-Resina, X., A brief history of theoretical chemistry, 2016, https://sites.google.com/a/r.umn.edu/prat-resina/divertimenti/a-brief-history-of-theoretical-chemistry (accessed December 15, 2016).

Vassberg, J.C., A brief history of FLO22, http://dept.ku.edu/~cfdku/JRV/Vassberg.pdf (accessed December 15, 2016).

of the development of selected major hardware advances, software and algorithm development, and scientific applications from a few fields. Looking at the first column in Table 1.1, one can see the tremendous growth in the power of the computers used in large-scale scientific computation. Advances in electronics and computer design have brought us from the ENIAC with 400 flops to Blue Waters with 13.34 petaflops, an increase in the maximum number of floating-point operations per second of more than 1015!

Tracing weather and climate modeling from von Neumann's first model on ENIAC, we can see that the computational power has allowed scientists to make rapid progress in the representation of weather and climate. In 1959, Lorenz laid the foundation for the mathematics behind weather events. By 1965, further advances in computing power and scientific knowledge provided the basis for the first global climate models. These have grown in scope to the present day to earth system models that couple atmospheric and ocean circulation that provide for the basis for the climate change forecasts of the international community.

Table 1.2 documents similar developments in computational chemistry, aeronautics and structures, and selected achievements in software and algorithms. The scientific advances were made possible not only by improvements in the hardware but also by the invention of programming languages, compilers, and the algorithms that are used to make the mathematical calculations underlying the models. As with weather modeling, one can trace the advancement of computational chemistry from the first simulation of molecules to the screening of drugs by modeling their binding to biomolecules. In aeronautics, the simulation of airflow over a wing in two dimensions has advanced to the three-dimensional simulation of a full airplane to create a final design. Similar timelines could be developed for every field of science and engineering from various aspects of physics and astronomy to earth and environmental science, to every aspect of engineering, and to economics and sociological modeling.

For those just getting introduced to these concepts, the terminology is daunting. The lesson at this point is to understand that computation has become an essential part of the design and discovery process across a wide range of scientific fields. Thus, it is essential that everyone understands the basic principles used in modeling and simulation, the mathematics underlying modeling efforts, and the tools of modeling along with their pitfalls.

1.2.1 Some Contemporary Examples

Although this book will not involve the use of large-scale models on supercomputers, some contemporary examples of large-scale simulations may provide insights into the need for the computational power described in Table 1.1. We provide four such examples.

Vogelsberger et al. created a model of galaxy formation comprised of 12 billion resolution elements showing the evolution of the universe from 12 million years after the Big Bang evolving over a period of 13.8 billion years (Vogelsberger et al., 2014). The simulation produced a large variety of galaxy shapes, luminosities, sizes, and colors that are similar to observed population. The simulation provided insights into the processes associated with galaxy formation. This example also illustrates how computation can be applied to a subject where experimentation is impossible but where simulation results can be compared with scientific observations.

Drug screening provides an example of how computer modeling can shorten the time to discovery. The drug screening pipeline requires a model of a target protein or macromolecular structure that is associated with a specific disease mechanism. A list of potential candidate compounds is then tested to see which have the highest affinity to bind to that protein, potentially inhibiting the medical problem. Biesiada et al. (2012) provide an excellent overview of the workflow associated with this process and the publically available software for accomplishing those tasks. The use of these tools allows researchers to screen thousands of compounds for their potential use as drugs. The candidate list can then be pared down to only a few compounds where expensive experimental testing is used.

The reports on global warming use comprehensive models of the earth's climate including components on the atmosphere and hydrosphere (ocean circulation and temperature, rainfall, polar ice caps) to forecast the long-term impacts on our climate and ecosystems (Pachauri and Meyer, 2014). The models:

> reproduce observed continental-scale surface temperature patterns and trends over many decades, including the more rapid warming since the mid-20th century and the cooling immediately following large volcanic eruptions (very high confidence) (IPC, 2013, p. 15).

Modeling and simulation has also become a key part of the process and designing, testing, and producing products and services. Where the building of physical prototypes or the completion of laboratory experiments may take weeks or months and cost millions of dollars, industry is instead creating virtual experiments that can be completed in a short time at greatly reduced costs. Proctor and Gamble uses computer modeling to improve many of its products. One example is the use of molecular modeling to test the interactions among surfactants in their cleaning products with a goal of producing products that are environmentally friendly and continue to perform as desired (Council on Competiveness, 2009).

Automobile manufacturers have substituted modeling for the building of physical prototypes of their cars to save time and money. The building of physical prototypes called *mules* is expensive, costing approximately $500,000 for each vehicle with 60 prototypes required before going into production (Mayne, 2005). The design of the 2005 Toyota Avalon required no mules at all—using computer modeling to design and test the car. Similarly, all of the automobile manufacturers are using modeling to reduce costs and get new products to market faster (Mayne, 2005).

These examples should illustrate the benefits of using modeling and simulation as part of the research, development, and design processes for scientists and engineers. Of course, students new to modeling and simulation cannot be expected to effectively use complex, large-scale simulation models on supercomputers at the outset of their modeling efforts. They must first understand the basic principles for creating, testing, and using models as well as some of the approaches to approximating physical reality in computer code. We begin to define those principles in Section 1.3 and continue through subsequent chapters.

1.3 THE MODELING PROCESS

Based on the examples discussed earlier, it should be clear that a model is an abstraction or simplification of a real-world object or phenomenon that helps us gain insights into the state or behavior of a complex system. Each of us creates informal, mental models all the time as an aid to making decisions. One example may be deciding on a travel route that gets us to several shopping locations faster or with the fewest traffic headaches. To do this, we analyze information from previous trips to make an informed decision about where there may be heavy traffic, construction, or other impediments to our trip.

Some of our first formal models were physical models. Those include simplified prototypes of objects used to evaluate their characteristics and behaviors. For example, auto manufacturers built clay models of new car designs to evaluate the styling and to test the design in wind-tunnel experiments.

Mississippi Basin Model
Vertical scale - 1:100; horizontal scale - 1:2000. Looking upstream on the Ohio River from Evansville. Indiana,
Tennessee, and Cumberland Rivers are in the foreground showing the site of the Kentucky and Barkley Dams.
Tradewater and Green Rivers are shown center. File No. 1270–4

FIGURE 1.1   Photo of portion of Mississippi River Basin model.

One of the most ambitious physical models ever built was a costly 200 acre model of the Mississippi River Basin used to simulate flooding in the watershed (U.S. Army Corps of Engineers, 2006). A photo of a portion of this model is shown in Figure 1.1. It included replicas of urban areas, the (Fatherree, 2006) stream bed, the underlying topography, levees, and other physical characteristics. Special materials were used to allow flood simulations to be tested and instrumented.

Through theory and experimentation, scientists and engineers also developed mathematical models representing aspects of physical behaviors. These became the basis of computer models by translating the mathematics into computer codes. Over time, mathematical models that started as very simplistic representations of complex systems have evolved into systems of equations that more closely approximate real-world phenomena such as the large-scale models discussed earlier in this chapter.

Creating, testing, and applying mathematical models using computation require an iterative process. The process starts with an initial set of simplifying assumptions and is followed by testing, alteration, and application of the model. Those steps are discussed in Section 1.3.1.

1.3.1 Steps in the Modeling Process

A great deal of work must be done before one can build a mathematical model on a computer. Figure 1.2 illustrates the steps in the modeling process. The first step is to analyze the problem and define the objectives of the model. This step should include a review of the literature to uncover previous research on the topic, experimental or field-measured data showing various states of the system and the measured outcomes, mathematical representations of the system derived from theories, and previous modeling efforts.

As that information is being gathered, it is also important to define the objectives of the modeling effort. There are several questions that should be addressed while considering the model objectives: What are the outcomes that we would like the model to predict? Are we interested in every possible outcome or is there a subset of conditions that would satisfy our model objectives? For example, we could be interested in just

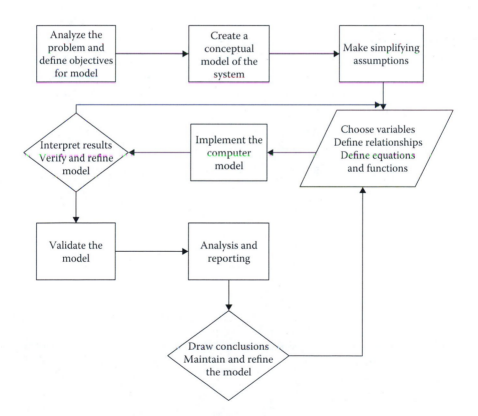

FIGURE 1.2   Major steps in the modeling process.

the *average* or *normal* state of affairs associated with a phenomenon or potential extreme events may be critical for our analysis. What level of accuracy is required for the predicted outcomes? This will impact the nature of the simplifying assumptions, input data, and computing algorithms that are required to build the model.

The second step in the process is to create a conceptual model of the system based on the analysis in the first step. A conceptual model will begin to specify all of the cause and effect relationships in the system, information on the data required and available to implement a model, and references to documents that were found in the initial analysis. The conceptual model should include a concept map showing the cause and effect relationships associated with the model and tables showing the different variables, data sources, and references. This can be done on a whiteboard, pencil and paper, or using a formal flowcharting or concept-mapping tool. There are several free tools for concept mapping. Cmap provides a free concept-mapping tool developed by the Florida Institute for Human and Machine Cognition. It creates nodes representing major components of a concept and labels the links between nodes with their relationships (Cmap, 2016). Mind Map Maker is a free mind-mapping tool provided as an app for Google Chrome users (Mindmapmaker, 2016). This tool allows one to create links between associated items. There are also a number of commercial packages in both categories.

Figures 1.3 and 1.4 are examples of a partially completed concept map and mind map showing the components of a model of the time it takes to make a car trip between two points.

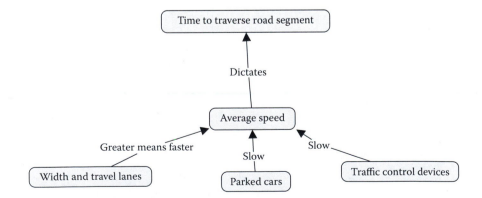

FIGURE 1.3   Partial concept map of model to calculate travel time using Cmap.

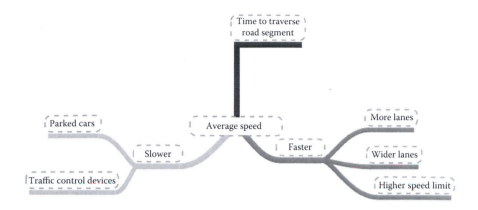

FIGURE 1.4   Partial mind map of model to calculate travel time using mind map maker.

The average speed across a road segment is slowed by parked cars and traffic control devices while wider lanes and higher speed limits take less time. The total time for a trip would need to add the average times associated with traversing each road segment. Thus, data on each segment will be needed as input to the model. Simple versions of such estimates are provided by global positioning satellite (GPS) equipment or the Internet mapping services that are available online. There are many other conditions that would impact this system. Modeling traffic conditions are a topic of one of the exercises at the end of the chapter.

Going back to Figure 1.2, one must choose which simplifying assumptions can be made in a model. This, in turn, leads to a selection of the data that would be needed, the variables that will drive the model, and the equations and mathematical functions that will comprise the model.

Once these items have been defined, a computer version of the model can be created and tested. The results must be verified to ascertain that the code is working properly. If the model is giving unexpected results with the code working properly, there may be a need to reexamine the simplifying assumptions and to reformulate the model. Thus, one may go through several iterations until the model is providing sufficiently accurate results. This can be validated against available experimental or field data to provide a quantitative assessment of model accuracy. Finally, the model can be used to undertake more detailed analysis and the results reported. As time goes on, the model must be maintained and may be improved by relaxing more of the assumptions and/or improving the input data. It should be noted that the judgment of whether a model is giving *reasonable* results is sometimes as much an art as a science. Confidence in that judgment is a function of the experience of the modeler and the breadth and depth of the previous research about the system under study. Of course the best validation of modeling results comes from comparisons with real data gathered from observations or experiments.

1.3.2 Mathematical Modeling Terminology and Approaches to Simulation

Similar to all scientific disciplines, mathematical modeling has its own unique vocabulary. Modeling novices may believe that the language used just creates a smoke screen that hides any problems associated with a model's development and use. Unfortunately, sometimes there is truth in that belief. Nevertheless, it is important to learn that language to enable a critical understanding of the modeling literature. We will begin with some basic definitions of modeling terms in this section.

It is also important to begin to understand the variety of approaches to modeling different types of systems. We will use some of the terminology we introduce to provide a few examples of different modeling approaches to simulate a variety of situations. We will then conclude this chapter with some exercises that let you delve deeper into the world of modeling and simulation.

1.3.3 Modeling and Simulation Terminology

By now, you should have your own concept of what constitutes a mathematical or computer model. A more formal definition is provided here.

> A mathematical model is a representation of a phenomenon or system that is used to provide insights and predictions about system behavior.

> Simulation is the application of a model to imitate the behavior of the system under a variety of circumstances.

There are several different ways to classify models. Models can be deterministic or probabilistic. Another term for probabilistic is stochastic meaning a random process or a process, which occurs by chance. A probabilistic model includes one or more elements that might occur by chance or at random while a deterministic model does not. A deterministic model applies a set of inputs or initial conditions and uses one or more equations to produce model outputs. The outputs of a deterministic model will be the same for each execution of the computer code with the same inputs. A probabilistic model will exhibit random effects that will produce different outputs for each model run.

Models can also be characterized as static or dynamic. A dynamic model considers the state of a system over time while a static model does not. For example, one could have a model of a material like a steel beam that considered its ability to bear weight without bending under a set of standard environmental conditions. This would be considered to be a static model of that system. A dynamic model of the same structure would simulate how the bearing strength and possible deformation of the beam would change under stresses over time such as under high temperatures, vibration, and chemical corrosion.

> A steady-state model is a model that has gone through a transient state such as a start-up or warm-up period and arrived at an observed behavior that remains constant.

An example of the steady-state model is the flow of fluid through a pipe. In the initial, transient state period, the pipe is empty and will fill with fluid under pressure until the capacity of the pipe is reached. This will be its steady-state condition. In economics, a steady-state economy is one that has reached a relatively stable size.

Perhaps making things more confusing, a dynamic model can have deterministic components. Such a model would track the state of a system over time and/or space. Given a current state, a deterministic function may be used to predict the future state of the system. Alternatively, the future state may be stochastic, which is impacted by random events.

Finally, dynamic models may be characterized as being discrete or continuous. A continuous model would represent time as a continuous function, whereas a discrete model divides time into small increments and calculates its state for each time period. In computer modeling, most (all?) dynamic models divide time into discrete increments to facilitate rapid calculations that mimic continuous systems.

1.3.4 Example Applications of Modeling and Simulation

In order to gain insights into system behavior, simulations are used to ask *what if* questions about how the system changes under different circumstances. How these questions are addressed depends in part on the type of model and its underlying mathematical structure. Solving those mathematical equations on a computer also leads to differences in programming logic or the algorithms that are used to calculate the most accurate answer most efficiently. We will discuss some of those algorithms as we go through the rest of this book. For now, it may help to provide some examples of different simulation approaches as they relate to various model types.

Deterministic models consist of one or more equations that characterize the behavior of a system. Most such models simplify the system by assuming that one or more causal variables or parameters are constant for a single calculation of the model outcomes.

For example, models of people's car trip behavior assume that the willingness to make a trip is inversely proportional to the trip distance. That is, people are more likely to make a trip from home to get to a destination that is closer than the one that is far away. Empirical studies have shown that this friction of distance changes depending on the nature of the trip. People are much more willing to make a longer trip to get to work than they are to do a convenience shopping trip. To simplify the system, these models assume a constant value of this friction of distance factor for each type of trip. When such a model is applied to a new urban area, there is some uncertainty that the constants found in previous studies in different places match the area where the model is being applied. Thus, a study is done where the model is run with different but reasonable variations in the constants to ascertain the impact of those changes on the predicted trips. Those can then be compared with a sample of real data to calibrate and validate the model.

Other examples of parametric studies include models of structures where different environmental conditions will alter system behavior, air and water pollution models where assumptions are made about the rate of dispersion of contaminants, and models of drug absorption into the blood stream where assumptions are made about absorption rates and excretion rates of the drug within the body. Many models include components that are both stochastic and deterministic where parametric studies are done on the deterministic components.

For dynamic models, the focus is on the behavior of the system over time and sometimes over space. For one group of such models called systems dynamics models, the state of the system at any time period is dependent, in part, on the state of the system at the previous time period. Simulations calculate the changes in the state of the system over time. An example is a model of ball being dropped from a bridge. As it is dropped the ball accelerates due to the force of gravity. At each time increment, the model will calculate the velocity of the ball and its position in space. That position will depend on where it was in the previous time period and how far it was dropped related to its velocity during that time period. The model will then predict when the ball will hit the water and at what velocity.

Stochastic models typically will have characteristics in common with dynamic models. The difference is that one or more of the governing parameters are probabilistic or could happen by random chance. One example is a model of the spread of a disease that is passed by human contact. A susceptible person may make contact with an infected person but will not necessarily become infected. There is a probability of being infected that is related to the virility of the disease, the state of health of the susceptible person, and the nature of the contact. A model of this system would simulate those probabilities to project the potential spread of a disease outbreak.

As we go through the rest of this book, we will describe the mathematical representation of each of these types of models and the programming steps needed to implement them on the computer. Exercises will involve the completion of example programs, the use of the model to make predictions, the analysis of model outcomes, and, in some cases, validation of model results. The exercises for this chapter focus on the modeling process and examples of how models have been used to solve research and production problems.

EXERCISES

1. Using a graphics program or one of the free concept-mapping or mind-mapping tools, create a complete conceptual map of the traffic model introduced earlier in the chapter. You should include all of the other factors you can think of that would contribute either to the increase or decrease in the traffic speed that might occur in a real situation.

2. Insert another concept mapping example here.

3. Read the executive summary of one of the following reports and be prepared to discuss it in class:

   a. PITAC report to the president

   b. Simulation-based engineering science report

   c. World Technology Evaluations Center

4. Using the student website for the book at http://www.intromodeling. com, choose an example model project in the document *example models for discovery and design* as assigned by your instructor. Read through the available material and then write a brief summary of the modeling effort and its characteristics using the summary template provided.

REFERENCES

Bartlett, B. N. 1990. The contributions of J.H. Wilkinson to numerical analysis. In *A History of Scientific Computing*, ed. S. G. Nash, pp. 17–30. New York: ACM Press.

Bell, G. 2015. Supercomputers: The amazing race. (A History of Supercomputing, 1960–2020). http://research.microsoft.com/en-us/um/people/gbell/MSR-TR-2015-2_Supercomputers-The_Amazing_Race_Bell.pdf (accessed December 15, 2016).

Bell, T. 2016. Supercomputer timeline, 2016. https://mason.gmu.edu/~tbell5/page2.html (accessed December 15, 2016).

Biesiada, J., A. Porollo, and J. Meller. 2012. On setting up and assessing docking simulations for virtual screening. In *Rational Drug Design: Methods and Protocols, Methods in Molecular Biology*, ed. Yi Zheng, pp. 1–16. New York: Springer Science and Business Media.

Cmap website. http://cmap.ihmc.us/ (accessed February 22, 2016).

Computer History Museum. 2017. Timeline of computer history, software and languages. http://www.computerhistory.org/timeline/software-languages/ (accessed January 2, 2017).

Council on Competitiveness. 2004. First Annual High Performance Computing Users Conference. http://www.compete.org/storage/images/uploads/File/PDF%20Files/2004%20HPC%2004%20Users%20Conference%20Final.pdf.

Council on Competitiveness. 2009. Procter & gamble's story of suds, soaps, simulations and supercomputers. http://www.compete.org/publications/all/1279 (accessed January 2, 2017).

Dorzolamide. 2016. https://en.wikipedia.org/wiki/Dorzolamide (accessed December 15, 2016).

Esterbrook, S. 2015. Timeline of climate modeling. https://prezi.com/pakaaiek3nol/timeline-of-climate-modeling/ (accessed December 15, 2016).

Fatherree, B. H. 2006. U.S. Army corps of engineers, Chapter 5 hydraulics research giant, 1949–1963, Part I: River modeling, potamology, and hydraulic structures. In *The First 75 Years: History of Hydraulics Engineering at the Waterways Experiment Station*, http://chl.erdc.usace.army.mil/Media/8/5/5/Chap5.htm. Vicksburg, MS: U.S. Army Engineer Research and Development Center (accessed October 15, 2016).

Goldstine, H. 1990. Remembrance of things past. In *A History of Scientific Computing*, ed. S. G. Nash, pp. 5–16. New York: ACM Press.

Gordon, S. I., K. Carey, and I. Vakalis. 2008. A shared, interinstitutional undergraduate minor program in computational science. *Computing in Science and Engineering*, 10(5): 12–16.

HPC University website. http://hpcuniversity.org/educators/competencies/ (accessed January 15, 2016).

International Panel on Climate Change. 2013. Climate change 2013—The physical science basis contribution of working Group I to the fifth assessment report of the IPCC, New York: Cambridge University Press. http://www.climatechange2013.org/images/report/WG1AR5_ALL_FINAL.pdf (accessed December 15, 2016).

Jameson, A. 2016. Computational fluid dynamics, past, present, and future. http://aero-comlab.stanford.edu/Papers/NASA_Presentation_20121030.pdf (accessed December 15, 2016).

Joseph, E., A. Snell, and C. Willard. 2004. Study of U.S. industrial HPC users. Council on Competitiveness. http://www.compete.org/publications/all/394 (accessed December 15, 2016).

Lax, P. D. 1982. Report of the panel on large scale computing in science and engineering. Report prepared under the sponsorship of the Department of Defense and the National Science Foundation. Washington, D.C.: National Science Foundation.

Lynch, P. and O. Lynch. 2008. Forecasts by PHONIAC. *Weather*, 63(11): 324–326.

Mayne, E. Automakers trade mules for computers, Detroit News, January 30, 2005, http://www.jamaicans.com/forums/showthread.php?1877-Automakers-Trade-Mules-For-Computers (accessed January 25, 2016).

McCartney, S. 1999. *ENIAC*. New York: Walker and Company.

Mindmapmaker website. http://mindmapmaker.org/ (accessed February 22, 2016).

National Science Foundation. 2006. Simulation-based engineering science: Report of the NSF blue ribbon panel on simulation-based engineering science. http://www.nsf.gov/pubs/reports/sbes_final_report.pdf.

Nick, T. 2014. A modern smartphone or a vintage supercomputer: Which is more powerful? http://www.phonearena.com/news/A-modern-smartphone-or-a-vintage-supercomputer-which-is-more-powerful_id57149 (accessed January 15, 2016).

Pachauri, R. K. and L. A. Meyer (ed.). 2014. Climate change 2014: Synthesis report. Contribution of working groups I, II and III to the fifth assessment report of the intergovernmental panel on climate change. http://www.ipcc.ch/report/ar5/syr/.

Prat-Resina, X. 2016. A brief history of theoretical chemistry. https://sites.google. com/a/r.umn.edu/prat-resina/divertimenti/a-brief-history-of-theoretical- chemistry (accessed December 15, 2016).

Reed, D. 2005. Computational science: America's competitive challenge. President's information technology advisory committee subcommittee on computational science. http://www.itrd.gov/pitac/meetings/2005/20050414/20050414_ reed.pdf.

Society of Industrial and Applied Mathematics (SIAM). 2001. Graduate education in computational science and engineering. *SIAM Review*, 43(1): 163–177.

Vassberg, J. C. A brief history of FLO22. http://dept.ku.edu/~cfdku/JRV/ Vassberg.pdf (accessed December 15, 2016).

Vogelsberger, M., S. Genel, V. Springel, et al. 2014. Properties of galaxies repro- duced by a hydrodynamic simulation. *Nature*, 509(8): 177–182. doi:10.1038 /nature13316.

Yasar, O. and R. H. Landau. 2001. Elements of computational science and engi- neering education. *SIAM Review*, 45(4): 787–805.

CHAPTER 2

Introduction to Programming Environments

2.1 THE MATLAB® PROGRAMMING ENVIRONMENT

MATLAB® (short for *matrix laboratory*) is a popular software package in many different science and engineering disciplines. It has a number of features that make it a good package for modeling and simulation. There are a large number of toolboxes available for license, as well as a number of community-provided toolboxes to solve common problems. In addition, as a fourth-generation programming language focused on numerical computing the language has built-in features that make it easy to work with vectors and matrices, allowing modelers to concentrate on their models, and not on implementing the details of a matrix operation.

2.1.1 The MATLAB® Interface

We will be using MATLAB R2016a in this book, as it is the most recent version available at the time of writing. The MathWorks has typically not made large changes to the user interface very frequently.

Once installed, when you launch MATLAB you will see the default interface as shown in Figure 2.1. Across the top is a ribbon toolbar with clearly marked functions, and tabs for *Home, Plots,* and *Apps.* We will return to few functions on the *Home* ribbon later in this chapter.

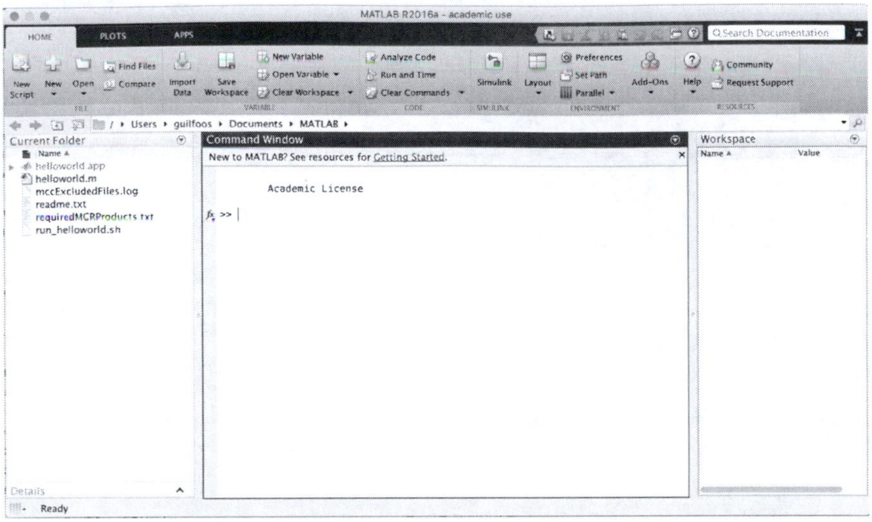

FIGURE 2.1   Default MATLAB® interface.

The large pane in the middle is the Command Window. This pane provides an interpreter and allows you to type MATLAB commands and see the results immediately.

The leftmost pane displays the current working directory's contents. You can browse through your directory tree using the widget directly above the three main panes. The current working directory is the first location where MATLAB will look for files when attempting to execute, open, or close files. There should be a *helloworld.m* file; if you double-click it the Editor will open, showing you the contents of the file and making it possible to edit and save the updated file. The Editor provides some rich tools, including syntax coloring, debugging, and more. When the Editor is open, a context-appropriate menu ribbon appears at the top, which includes debugging controls. We will return to those later in this chapter.

The pane to the far right is the Workspace, which displays the variables currently being used by MATLAB, and their value. Double-clicking on a variable in this list will open the Variables pane, which allows fuller inspection and editing of variables, including large matrices. Returning to the menu ribbon at the top of the window, you should notice *Import Data* and *Save Workspace*, which allow you to quickly import and export datasets to and from your Workspace.

2.1.2 Basic Syntax

MATLAB (the application shares a name with its programming language) is a relatively flexible language that uses certain characters for flow control such as "{" and "}". Also, unlike a language like C, which requires a special termination character at the end of every line (a semicolon), that character is *optional* in MATLAB. However, the decision to include it or not is important. Including a semicolon at the end of a line will *suppress the display of output* related to the execution of that line. Omitting the semicolon will tell the interpreter to display the result of that command. We will see examples of this later. We will introduce various concepts in a just-in-time basis as we work through the course materials. As we go through a number of the syntax examples later, we suggest you to try them out in the Command Window, Workspace, or Editor, as appropriate.

2.1.2.1 Variables and Operators

All programming languages provide variables—a method to store and manipulate data that may be different from one run of the program to the next—and ways to manipulate those variables. We will introduce some basics here and will leave some of the more advanced tools until later chapters.

In the most basic sense, a variable is just a name that we use to refer to a value that may change over time. In MATLAB, we can simply create variables as they are needed, without having to declare a variable type (as is required in many other languages). For example:

```
x = 2
five = 5
z = 3.14159
my_string = 'Hello World!'
```

Type these commands into the console. Once a variable is declared, it can be recalled and used in an appropriate calculation. As an example enter this:

```
y=x*five;
```

You will note that we included a semicolon, which suppressed the output of the command, unlike the previous examples. To see the result, type:

```
y
```

Now that you have some variable in memory, you can look at the Workspace to see the list of variables and their current values.

Variables have something called *scope*, which defines where they are visible. For example, variables defined in the Command Window are *global*; they can be referenced, used, and modified from any other piece of code. For example, if we define "x" in the Command Window, and then in my_script.m if we add the command "x", when we run that program, it will print the value of "x".

However, variables defined within a function are only visible within that function. These are called *local variables*. We will come back to this idea later when you begin to create full programs.

You can delete variables in two ways. The first is the *clear* command:

```
clear x
```

The second is to go to the Workspace window, right-click on the variable you wish to delete, and select *Delete* from the pop-up menu.

You will also note that the Workspace gives you tools such as *rename* and *edit*, which will be useful.

In computational science, we call variables that hold a single value a *scalar*. This is slightly different than the mathematical definition of *scalar*. MATLAB supports a number of arithmetic operations (Table 2.1).

You can try some of these calculations in the Command Window by typing in the left side of the equation. That will be defined as input, and the results will emerge as output.

Please note that, in MATLAB, some of these operators are also used for matrix operations, and MATLAB will return an error (or do something you may not be expecting) if you are using matrices. You can explicitly tell the interpreter that you want to do scalar operations by prepending the operators with a period ("."). For example, to conduct a scalar multiplication instead of a matrix multiplication, use ".*". This can be a concern

TABLE 2.1   MATLAB® Mathematical Operators

Symbol	Operation	Example
+	Addition	$2 + 2 = 4$
−	Subtraction	$4 - 1 = 3$
/	Division	$9/3 = 3$
*	Multiplication	$8 * 6 = 48$
^	Exponential	$3 \wedge 2 = 9$

TABLE 2.2   Order of Execution for
Mathematical Operations in MATLAB®

()	Items enclosed in parentheses
^	Exponentiation
*, /	Multiplication, division
+, −	Addition and subtraction
=	Assignment

when doing multiplication, division, and exponentiation. You can include the "." safely at any time you want to perform the scalar operation.

MATLAB follows the normally expected order of operations: exponents and roots, followed by multiplication and division, followed by addition and subtraction. Specifically, operations should be carried out in the following order, and for operations at the same level from left to right as shown in Table 2.2.

For example,

```
(3 * 2) ** 3 + 6
```

What happens if we execute this command without the parentheses?

Please note that we have not included all operators. Comparison and logical operators will be discussed later.

2.1.2.2 Keywords

MATLAB reserves certain words, called *keywords*, which cannot be used as variable names. Note that these words will be colored in blue when they are typed in the Editor or Command Window. They are listed in Table 2.3.

TABLE 2.3   MATLAB® Reserved Keywords

break	case
catch	classdef
continue	else
elseif	end
for	function
global	if
otherwise	parfor
persistent	return
spmd	switch
try	while

An additional word of warning: It is possible to overwrite a function name (but not a keyword) with a variable name (and vice versa). Use care when selecting variable names, or you might experience unexpected errors when executing code.

2.1.2.3 Lists and Arrays

In programming, we often have a group of homogeneous variables that represent multiple values of either inputs or outputs from a model. For example, as inputs we might have traffic counts for a particular location for different hours and/or different days, multiple values of an environmental indicator such as air pollution for different times, or a sequence of values of a model parameter we will use to test its impacts on the model outcomes. We then would want to store these results in a similar group.

One way to represent such a group of variables in MATLAB is called an array. Interestingly, in MATLAB *everything* is an array, including the scalar values we were working with before. An array is defined to include places in memory for multiple items indexed by a sequence number. It can be declared in several ways:

```
traffic=[200,150,350,235,450];
nox = zeros(1,365);
```

If you type in the first example, you will see that a traffic variable is created with five items with the values indicated. In the second example (our first use of a function), an array is created with 365 items initially set to a value of 0. This could be, for example, space to hold the average daily nitrogen oxide content in the air. Notice how they are represented in the Workspace.

We can use the index to operate on each of the items in the list in turn or can operate on any individual item by using its index number. To see a single value, we use the variable name with the index in parentheses. Try this and see what happens:

```
traffic(2)
```

You should get the value 150. This is because MATLAB starts all lists and arrays with the index number 1. So if you put traffic(1) in the console, you should get 200. What happens if you put this in:

```
traffic
```

A one-dimensional array is often called a vector, whereas a two-dimensional array is called a matrix. Arrays can use matrix mathematics to operate on the entire array in specific ways we will introduce later.

We have already seen the basic interface for creating vectors and matrices.

```
x=[1, 2, 3, 4];
y=[1, 2; 3, 4];
```

Note that we created a 1 × 4 vector and a 2 × 2 matrix with the previous commands. Examine how these variables are characterized in the Workspace and how they are shown in the Command Window when you query for the contents of x and y.

There are several special functions to create certain types of arrays. An array with all zeros can be created with the "zeros()" function, as we saw earlier, whereas an array of ones can be created with the "ones()" function. An identity array can be created with the "eye()" function. Try these functions:

```
x=zeros(6)
y=ones(8)
```

To create an evenly spaced array, MATLAB provides a function called "colon()", which also has shorthand using the ":" operator. To create an array that goes from 0 to 5 (inclusive—so 6 elements), you can type "0:5". For an array that starts with 2 and ends with 8, you can type "2:8" or "colon(2,8)".

```
z=2:8
```

You can also specify a custom step size, instead of being restricted to "1" as the default, by adding a third parameter, located between the start and stop points. For example, 3:.2:4 will return an array containing [3 3.2 3.4 3.6 3.8 4].

Another useful function for creating vectors containing regularly spaced values is "linspace()". Rather than specifying the size of the step, you specify the number of elements you want in your array. For example, linspace(3,4,6) will return an array containing [3. 3.2 3.4 3.6 3.8 4].

```
myarray=linspace(3,4,6)
```

We can access (and modify) individual elements in arrays in MATLAB. Use the linspace() example above to create the array "myarray". We can look at the second element of myarray by typing "myarray(2)" in the Command Window. We can change the value of the second element in "myarray" simply by using it on the left side of an assignment expression.

```
myarray(2) = 2
```

This will turn "myarray" into [3 2 3.4 3.6 3.8 4].

We can also extract portions of arrays called *slices*. For example, we could create a slice of "myarray" containing the second and third elements with the command "myarray(2:3)." The argument is in the form "start:step:stop" (the same as the "colon()" function). The *step* argument is optional (and assumed to be 1 unless specified), and *stop* can use the keyword *end* to mean *go to the end of the list*. "myarray(1:3)" will return [3 2 3.4], whereas "myarray(4:end)" will return [3.6 3.8 4]. "myarray(1:2:end)" will select every other element in the array: [3 3.4 3.8].

2.1.3 Common Functions

We could not possibly exhaustively list every function included in MATLAB here, much less everything available in the numerous toolboxes. Throughout the book, we will introduce additional functions as needed, but Table 2.4 shows a few important ones available in the base MATLAB program. To use these functions, enclose the target variable inside the parentheses.

2.1.4 Program Execution

One of the things that makes MATLAB a powerful tool is that you can both work interactively in the Command Window, and you can also write

TABLE 2.4 Example Built-in Functions for MATLAB®

abs()	Absolute value.
mod()	Take two noncomplex numbers and return their remainder when using long division.
whos()	Returns the variables in the current Workspace. "whos GLOBAL" returns the globally scoped variables.
size()	Returns the size of the array, meaning the number of items in each dimension.
max()	Returns the largest item in the array.
min()	Returns the smallest item in the array.
open()	Opens a file. We will talk about file input/output in more detail later.
disp()	Prints arrays to the display, while not showing the variable name. Useful for interacting with users of your programs.

complicated programs capturing very detailed work flows to ensure accurate repeatability. Programs can be executed outside of the Command Window by passing the program file (in this example called *myprogram.m*) to the MATLAB executable in your operating system's command window:

```
matlab -r myprogram.m
```

We can also run these programs inside of the Command Window or via the MATLAB graphical user interface, which provides some additional debugging capabilities.

2.1.5 Creating Repeatable Code

In MATLAB, creating repeatable code is as simple as typing the commands you want into a single file with a ".m" extension. One common method for code development in interpreted languages is to interactively manipulate variables until you begin to see how to get the results you want, and then pull those commands from your history and put them into the program file. Later we will explore flow control—how to execute some blocks of code but not others—and how to create your own functions or classes, but for now, we will focus on a simple list of commands to execute.

MATLAB starts off with a file called "helloworld.m" in the default working directory. You can directly edit this file to explore this functionality (double-click on it to open it in the Editor), but you can also use *New* button on the menu ribbon to create other files. Let us edit helloworld.m to modify the basic *Hello World* program. Add a line at the bottom to display (*disp*) the string "My name is 'Hal'.". Your Editor pane should look similar to Figure 2.2.

You can execute this program by clicking the gray triangle labeled *Run* directly above the Editor, which will execute the program in the Command Window. You should see a *helloworld* command, followed

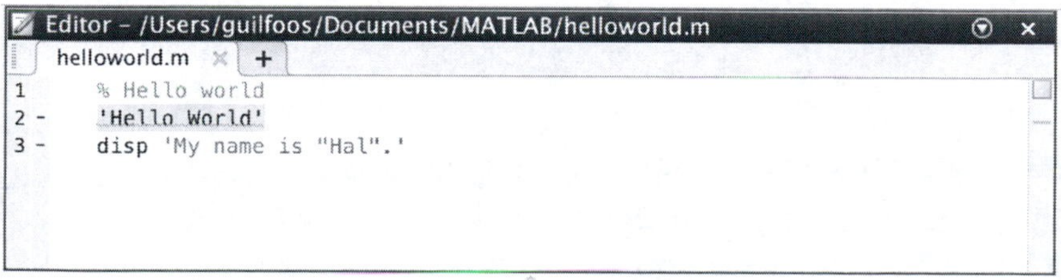

```
 Editor – /Users/guilfoos/Documents/MATLAB/helloworld.m
 helloworld.m  ×  +
1     % Hello world
2 -   'Hello World'
3 -   disp 'My name is "Hal".'
```

FIGURE 2.2   MATLAB® Hello World script.

by your code output on the next line. Alternatively, you can simply type *helloworld* in your Command Window.

2.1.6 Debugging

MATLAB includes some debugging tools, which can be very useful in discovering problems in your code.

If you set a *breakpoint* in your code, simply running the code will trigger the debugger when the breakpoint is reached. Breakpoints allow you to target a specific point in the code that you wish to stop and investigate. When the debugger reaches a breakpoint, you can either run commands in the Command Window to investigate the state of your program, or you can use the Workspace and Variables panes to look at your data. You can set breakpoints by clicking on the "-" in the gray space to the left of the line you wish to stop execution at in the Editor.

The most common bugs you are likely to encounter when working on modeling and simulation problems will be when your data do not contain the values you expected when executing a certain block of code, and MATLAB's debugger is a useful tool for discovering this.

Once you wish to continue execution, you can *Step* through the program one line at a time, *Step In* to the function of the current line, *Step Out* and run until the current function ends, *Continue* until the debugger hits the next breakpoint, or *Quit Debugging*. You will need to exit the debugger to return the console to the normal mode. The exercises include an example where you can try out the debugger.

2.2 THE PYTHON ENVIRONMENT

Python is a very popular high-level general-purpose programming language. As an interpreted language, code does not need to be compiled and can be run on any system where an interpreter is available. The core language is Free and Open-Source Software (FOSS), and can be acquired and used at no cost. There is a very large library of routines available, including ones specifically designed for scientific computing, such as SciPy or NumPy.

2.2.1 Recommendations and Installation

There are a number of integrated development environments (IDEs) available for Python, many of which would be quite good for this course. We are recommending the use of Spyder, a free IDE, which includes a number of MATLAB-like features that make it especially useful for modeling and simulation use. In addition, you will need to download and install an interpreter; we recommend Anaconda, which is available for Mac, Linux, and Windows at https://www.continuum.io/downloads and which provides a bundle of scientific programming packages in the base installation (including Spyder). We recommend getting Python 3; there are some minor syntax differences between Python 2 and Python 3, and Python 3 has a few new features that make it better suited to our purposes. This book uses Python 3 syntax.

2.2.2 The Spyder Interface

The Spyder IDE is already included in the Anaconda installation. If you are not using Anaconda, you can find Spyder and installation instructions at https://pythonhosted.org/spyder/.

To launch Spyder, you can select it from the list of programs for your computer or you can run *Spyder* in a terminal or command window. When you start Spyder, you should see an interface with multiple panes as shown in Figure 2.3. In the lower right of the window, Spyder provides an interactive console. By default, it launches *IPython*, which is an enhanced interactive Python terminal that provides some improvements that make it more useful for interactive programming than the standard Python interpreter. Here, you can type Python commands, which will be executed immediately.

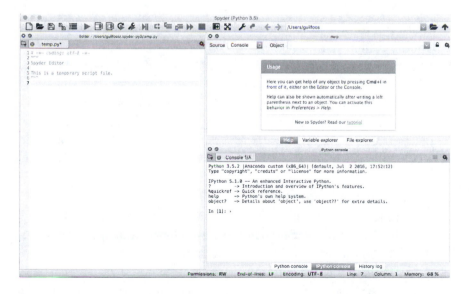

FIGURE 2.3   Default Spyder interface.

The upper right region of the window shows the Help pane. It will automatically show documentation for a function or object being instantiated in the editor or console, or you can type a function name in the *Object* field at the top of the window to pull up the corresponding documentation.

In the same pane as the Help is the Variable Explorer. Clicking on the *Variable Explorer* button at the bottom of the pane will switch the view to the Variable Explorer. This gives you a view into the global variables for the current console. It will be empty until you enter commands to create or modify variables in the console or run a script in the editor. Most common variable types can be displayed, and variables can be edited in this interface. You can do quick plots of arrays and show matrices as images.

You can also save sets of variables in a Spyder-specific format, MATLAB data format, or HDF5 format by using the Disk icon to the right of the pane. In addition, you can import data from a large variety of supported formats using the Import Data icon (downward pointing arrow icon).

The left pane shows the Spyder editor. It enables editing of Python (and other languages!) and includes features such as syntax coloring, function/class/method browsing, code analysis, introspection, and more. At the top of the Spyder window are debugging controls that allow you to execute the Editor pane's current file in the IPython console; we will cover those in more detail later.

2.2.3 Basic Syntax

Python uses whitespace to designate code blocks. In short, this means indentation is how we group lines of code together for flow control. This will become clearer later. Also, unlike many other languages (including MATLAB), semicolons are not used as line terminators. We will fill in other details in a just-in-time basis as we work through the course materials. As we go through a number of the syntax examples below, we suggest you to try them out in the Spyder interface by inserting them in console, help, or editor as appropriate.

2.2.3.1 Variables and Operators

All programming languages provide variables—a method to store and manipulate data that may be different from one run of the program to the next—and ways to manipulate those variables. We will introduce some basics here, and leave some of the more advanced tools until later chapters.

In the most basic sense, a variable is just a name that we use to refer to a value that may change over time. In Python, we can simply create variables as they are needed, without having to declare a variable type (as is required in many other languages). For example:

```
x = 2
five = 5
z = 3.14159
my_string = "Hello World!"
```

Type these commands into the console. Once a variable is declared, it can be recalled and used in an appropriate calculation. As an example, enter this:

```
y=x*five
```

To see the result, type:

```
y
```

Now that you have some in memory, open the variable explorer to see the list of variables and their current values.

Variables have something called *scope*, which defines where they are visible. For example, variables defined in the console are *global*; they can be referenced, used, and modified from any other piece of code. For example, if we define "x" in the console, and then in temp.py if we add the command "print(x)", when we run that program, it will print the value of "x".

However, variables defined within a code block (such as a function or an indented section of code) are only visible within that code block, and any dependent blocks (subblocks). These are called *local variables*. We will come back to this idea later when you begin to create full programs.

You can delete variables in two ways. The first is the *del* command:

```
del x
```

The second is to go to the Variable Explorer, right-click on the variable you wish to delete, and select *Remove* from the pop-up menu.

You will also note that the Variable Explorer gives you tools such as *rename* and *edit* that will be useful.

TABLE 2.5 Python Mathematical Operators

Symbol	Operation	Example
+	Addition	$2 + 2 = 4$
−	Subtraction	$4 − 1 = 3$
/	Division	$9/3 = 3$
%	Modulo	$7 \% 2 = 1$
*	Multiplication	$8 * 6 = 48$
//	Floor division	$7//2 = 3$
**	Exponential	$3 ** 2 = 9$

TABLE 2.6 Order of Execution for Mathematical Operations in Python

( )	Items enclosed in parentheses
**	Exponentiation
*, /, %, //	Multiplication, division, modulo, floor division
+, −	Addition and subtraction
=	Assignment

In computational science, we call variables that hold a single value a *scalar*. This is slightly different than the mathematical definition of *scalar*. Python supports a number of arithmetic operations, as shown in Table 2.5.

You can try some of these calculations in the console by typing in the left side of the equation. That will be defined as input, and the results will emerge as output.

Python follows the normally expected order of operations: exponents and roots, followed by multiplication and division, followed by addition and subtraction. Specifically, operations should be carried out in the following order, and for operations at the same level from left to right as shown in Table 2.6.

For example,

```
(3 * 2 % 4 // 2) ** 3 + 6
```

What happens if we execute this command without the parentheses?

Please note that we have not included all operators. Comparison and logical operators will be discussed later.

2.2.3.2 Keywords

Python reserves certain words, called *keywords*, which cannot be used as variable names. Note that Spyder will color these words in magenta when they are typed in the Editor or console. They are listed in Table 2.7.

TABLE 2.7   Python Reserved Keywords

and	import
as	in
assert	is
break	lambda
class	none
continue	nonlocal
def	not
del	or
elif	pass
else	raise
except	return
false	true
finally	try
for	while
from	with
global	yield
if	

An additional word of warning: It is possible to overwrite a function name with a variable name (and vice versa). Use care when selecting variable names, or you might experience unexpected errors when executing code.

2.2.3.3 Lists and Arrays

In programming, we often have a group of homogeneous variables that represent multiple values of either inputs or outputs from a model. For example, as inputs we might have traffic counts for a particular location for different hours and/or different days, multiple values of an environmental indicator such as air pollution for different times, or a sequence of values of a model parameter we will use to test its impacts on the model outcomes. We then would want to store these results in a similar group.

One way to represent such a group of variables in Python is called a list. A list is defined to include places in memory for multiple items indexed by a sequence number. It can be declared in several ways:

```
traffic=[200,150,350,235,450]
nox = [0.]*365
```

If you type in the first example, you will see that a traffic variable is created with five items with the values indicated. In the second example, a list is created with 365 items initially set to a value of 0. This could be, for example, space to hold the average daily nitrogen oxide content in the air. Notice how they are represented in the Variable Explorer.

We can use the index to operate on each of the items in the list in turn or can operate on any individual item by using its index number. To see a single value, we use the variable name with the index in brackets. Try this and see what happens:

```
traffic[1]
```

You should get the value 150. This is because Python starts all lists and arrays with the index number 0. So if you put traffic[0] in the console, you should get 200. What happens if you put this in:

```
traffic
```

For many programming applications, we will need an array of values that have a more specific mathematical definition. An array is an ordered sequence of values much like a list. A one-dimensional array is often called a vector, whereas a two-dimensional array is called a matrix. Unlike a list, numerical arrays can use matrix mathematics to operate on the entire array in specific ways we will introduce later. For now, we just wish to define how an array is declared in Python. To do so, we need to import a special module into our Python environment called NumPy. We can import NumPy into our code and rename it as *np* with the following command:

```
import numpy as np
```

For the rest of this textbook, when using functions or objects provided by NumPy, we will assume that it has been imported as *np*, which is consistent with the documentation for NumPy. Please note that renaming a package upon import does *not* rename the package in the Help.

The basic interface for creating vectors and matrices is NumPy's "array()" function, and it can accept a list object as input.

```
x=np.array([1, 2, 3, 4])
y=np.array([[1, 2], [3, 4]])
```

Note that we created a 1 × 4 vector and a 2 × 2 matrix with the previous commands. The data type created by array() is *ndarray*, which is distinct from a list, and has some supporting functions that lists do not have. Examine how these variables are characterized in the variable explorer and how they are shown in the console when you query for the contents of x and y.

As with lists, Python arrays are what we call *zero-indexed*. What this means is that the very first element in the array (in any dimension) is in the position labeled *zero*. This is a very common convention in many programming languages, compared with the everyday convention of calling the first element *index 1*. This will be important when iterating over arrays or selecting individual elements or ranges.

The array() function has a number of optional parameters, which we will not cover in detail. There are several special NumPy functions to create certain types of arrays. An array with all zeros can be created with the "zeros()" function, whereas an array of ones can be created with the "ones()" function. An identity array can be created with the "eye()" function. Try these functions:

```
x=np.zeros(6)
y=np.ones(8)
```

One NumPy function we will use heavily will be "arange()". In its most simple form, it will give you a vector starting at zero and counting by one to the end point you specify. For example, arange(5) will return an *ndarray* containing [0, 1, 2, 3, 4]. Alternatively, you can supply optional arguments to specify the start point and the size of the step. For example, arange(3,4,.2) will return an *ndarray* containing [3., 3.2, 3.4, 3.6, 3.8].

```
z=np.arange(6)
```

Another useful NumPy function for creating vectors containing regularly spaced values is "linspace()". Rather than specifying the size of the step, you specify the number of elements you want in your *ndarray*. For example, linspace(3,4,6) will return an *ndarray* containing [3., 3.2, 3.4, 3.6, 3.8, 4.].

```
myarray=np.linspace(3,4,6)
```

We can access (and modify) individual elements in arrays in Python. Use the linspace() example above to create the array "myarray". We can look at the second element of myarray by typing "myarray[1]" in the console.

Remember that previously we explained that Python is *zero-indexed*, meaning myspace[0] refers to the first element, and myspace[1] refers to the second. We can change the value of the second element in "myarray" simply by using it on the left side of an assignment expression.

```
myarray[1] = 2
```

This will turn "myarray" into [3., 2., 3.4, 3.6, 3.8, 4.].

We can also extract portions of arrays called *slices*. For example, we could create a slice of "myarray" containing the second and third elements with the command "myarray[1:3]". The argument is in the form "start:stop:step", and it does *not* include the stop element. The "step" argument is option (and assumed to be 1 unless specified), whereas start and stop, if not specified, are assumed to be the beginning and end of the array. "myarray[:3]" will return [3., 2., 3.4], whereas "myarray[3:]" will return [3.6, 3.8, 4.]. "myarray[::2]" will select every other element in the array, starting with position zero: [3., 3.4, 3.8].

2.2.4 Loading Libraries

Code libraries are bundled up in things called *modules*. To make those modules available in our programs, we need to *import* them as we did for the NumPy module earlier. For example, a module called *math* provides a number of basic mathematics functions. To load that module, we issue this command in our script or IPython console:

```
import math
```

Now, we can execute functions from the *math* module. For example, we can find the square root of a value:

```
math.sqrt(9)
```

Go ahead and try out these commands in the console.

Importing a module is not persistent; when you restart a console, you will need to reimport any modules. Remember, you can find documentation for modules and functions in the Help. Now that you have imported the math module, type the term math.sqrt in the Help pane to get the documentation for this command. You can also put your cursor on the line in the console where you typed math.sqrt and hit the Control and I key to get the same object feedback.

Rather than importing an entire module, we can also import just a single function of that module. This has the advantage of simplifying the call to the function later in your code. Returning to the square root function:

```
from math import sqrt
```

Now, you can directly execute sqrt:

```
sqrt(9)
```

However, other functions in the *math* module will not be available.

2.2.5 Common Functions

We could not possibly exhaustively list every function included in the common modules here, much less everything available in the Anaconda distribution. Throughout the book, we will introduce additional functions as needed, but Table 2.8 shows a few important ones from the *builtins* module (available without loading). To use these functions, enclose the target variable inside the parentheses.

TABLE 2.8   Example Built-in Functions for Python

abs()	Absolute value.
divmod()	Take two noncomplex numbers and return their quotient and remainder when using long division.
float()	Turns the input string or number into a floating-point number.
globals()	Returns a variable called a *dictionary* that shows the current global symbol table. This includes everything you can see in the Variable Explorer, but also things like loaded modules and command history.
int()	Turns the input string or number into an integer.
len()	Returns the number of items in the input. Accepts strings, lists, dictionaries, sets, etc.
max()	Returns the largest item in the input list.
min()	Returns the smallest item in the input list.
open()	Opens a file. We will talk about file input/output in more detail later.
print()	Prints objects, either to a specific file or to the display. Useful for interacting with users of your programs.
range()	Creates lists of arithmetic progressions. Will be very useful; read the documentation in the Help for more details. We will visit a very similar function later in this chapter.

2.2.6 Program Execution

One of the things that makes Python a powerful tool is that you can both work interactively in the IPython console, and you can also write complicated programs capturing very detailed work flows to ensure accurate repeatability. Programs can be executed outside of the Spyder environment by passing the program file (in this example called *myprogram.py*) to the python interpreter:

```
python myprogram.py
```

We can also run these programs inside of the Spyder environment, which provides some additional debugging capabilities.

2.2.7 Creating Repeatable Code

In Python, creating repeatable code is as simple as typing the commands you want to execute into a single file with a ".py" extension. One common method for code development in interpreted languages is to interactively manipulate variables until you begin to see how to get the results you want, and then pull those commands from your history and put them into the program file. Later we will explore flow control—how to execute some blocks of code but not others—and how to create your own functions, classes, or modules, but for now we will focus on a simple list of commands to execute.

The Spyder Editor starts off with a file called "temp.py" opened. You can directly edit this file to explore this functionality, but you can also use *New File* under the *File* menu to create other files. Let us edit temp.py to create the basic "Hello World" program. Add a line at the bottom to *print* the string "Hello World!". Your Editor pane should look similar to Figure 2.4.

You can execute this program by clicking the "Run File" button (looks like a "Play" button) directly above the Editor, which will execute the program in the IPython console. You should see a "runfile()" command created by the IDE, followed by your code output on the next line.

Here is an Easter Egg: type "import __hello__" in the console. (Note: "__" is a repeated underscore character.)

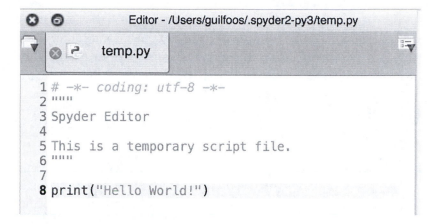

FIGURE 2.4   Python Hello World script.

2.2.8 Debugging

One advantage to Spyder over testing code directly in the interpreter as described above is that Spyder includes some debugging tools, which can be very useful in discovering problems in your code.

To launch the debugger, click on the "Debug File" button, which shows a *play* and *pause* button next to each other. (Hovering your mouse over any button in the Spyder toolbar will display a tip in the lower left of the window explaining what the button does.) Breakpoints allow you to target a specific point in the code that you wish to stop and investigate. When the debugger reaches a breakpoint, you can either run commands in the console to investigate the state of your program, or you can use the Variable Explorer to look at your data. You can set breakpoints by double-clicking in the gray space to the left of the line you wish to stop execution at; the debugger will run to that line by default or will stop at the very first line otherwise.

The most common bugs you are likely to encounter when working on modeling and simulation problems will be when your data does not contain the values you expected when executing a certain block of code, and Spyder's debugger is a useful tool for discovering this.

Once you wish to continue execution, you can (examining the debugging buttons from left to right) step through the program one line at a time, step into the function of the current line, run until the current function ends, run until the debugger hits the next breakpoint, or exit. You will need to exit the debugger to return the console to the normal mode. The exercises include an example where you can try out the debugger.

EXERCISES

1. Basic Arithmetic

   Convert the following equations into valid MATLAB or Python commands, and submit both the input and output of the command from the interpreter.

   a. $(2+3) \times 7$

   b. $\dfrac{3^3}{2}$

   c. $\dfrac{12 \times 3.3^4}{1 + 2.7^{4.6}}$

2. Creating and Editing Matrices

   Create the following matrices, and submit both the input and output of the command from the interpreter.

   a. $\begin{bmatrix} 4 & 6 & 2 \end{bmatrix}$

   b. $\begin{bmatrix} 3.5 & 3.14 \\ 7 & 0.25 \end{bmatrix}$

   c. $\begin{bmatrix} 1 & 0 & 0 \\ 0 & 1 & 0 \\ 0 & 0 & 1 \end{bmatrix}$

   d. $\begin{bmatrix} 1.3 & 1.6 & 1.9 & 2.2 & 2.5 \end{bmatrix}$

   For the following problems, create the matrices as described, and then edit as directed, and submit the editing command and the output from the interpreter.

   e. Create the array $\begin{bmatrix} 1 & 2 & 3 & 4 & 5 \end{bmatrix}$, and replace the third element with the number 9.

f. Create the array $[3 \quad 6 \quad 2 \quad 9 \quad 12]$, and replace the second element with the value in that location cubed. That is, cube the value found in that location and store it back in the same location.

g. Create the identity matrix $\begin{bmatrix} 1 & 0 & 0 \\ 0 & 1 & 0 \\ 0 & 0 & 1 \end{bmatrix}$ and replace the element

at location 2,3 with the number of elements in the identity matrix.

3. Saving and loading data

Use the materials on the book website to download the dataset necessary for these problems in either MATLAB or Python.

a. Load the Chapter 2 dataset (Chapter2.m or Chapter2.py). What is the value in variable "x"?

b. Clean out the variables in your Workspace or Variable Explorer. Load the Chapter 2 dataset. Modify the variable "z" to equal $x \times y$. Save the dataset, and submit.

4. Debugging

Use the materials on the book website to download the sample code necessary for these problems in either MATLAB or Python.

a. Open the code in the Editor. Set a breakpoint on line 22. Run the code in the debugger, and report the value of the variable "x" when the code hits the breakpoint.

b. Open the code in the Editor. Set a breakpoint on line 19 and on line 24. Run the code in the debugger, and report the value of the variable "y" every time the code hits the first breakpoint. Then report the value of "z" when the code hits the second breakpoint.

CHAPTER 3

Deterministic Linear Models

3.1 SELECTING A MATHEMATICAL REPRESENTATION FOR A MODEL

Each of the model examples presented in Chapter 1 is represented by one or more mathematical equations whose goal is to accurately mimic the behavior of the real system. The logical question that arises to the modeling neophyte is how does one choose the appropriate set of equations? This is where computer modeling is closely tied to the traditional approaches to the advancement of science—theory and experimentation.

The initial approach to understanding the world around us was for people to observe and experiment. By recording the observations of phenomena over time and space, it is possible to begin to recognize patterns in changes and the possible underlying causes of those changes. Experiments where the underlying conditions are controlled provide further insights into the relationships between the observed effects and their possible causes. Those observations can then be used to represent the changes in mathematical terms.

Over time the level of knowledge related to a particular type of phenomenon increases allowing scientists to formulate theories about the relationships that go beyond the recorded data. The ability to confirm a particular theory is then tied to our ability to accurately observe and measure the system through additional experimentation and simulation.

In this way, some of our theories become the *laws* we learn about in studying science and engineering.

The most appropriate mathematical model of a phenomenon is not always the most complex or sophisticated approach. The approach we use is tied to the purpose of the modeling effort and the required degree of accuracy associated with the use of the model. For example, we might be interested in the impacts of adding new manufacturing facilities on the quality of the air in a particular region for long-range planning purposes. The initial model of air quality might consider the local air circulation conditions and the existing levels of pollution but only make a general forecast of a *typical* manufacturing facility and its air emissions to determine roughly how much additional industry, if any, could be added without causing major air quality problems. This simple model would make many assumptions about the nature of the emissions from this typical plant in the context of a simplified range of weather conditions. This type of approach is often called a *screening* model, meaning it is sacrificing accuracy to allow a rapid determination of the general nature of the situation that can be done without a large investment in data collection and model development.

On the other hand, if a particular manufacturing facility is proposed in the same region, we will want to use a more sophisticated air quality model that takes into account the nature of the emissions that the plant will make, the potential reductions in emissions that can be achieved with pollution control equipment, and a much more refined look at the air circulation in the region to determine whether the resulting air quality will cause harm to human health and the environment.

This is the first of a series of chapters where we will provide examples of models with differing mathematical representations. These will include deterministic models with linear and nonlinear representations, probabilistic models, static models, and dynamic models where changes occur over time and/or space. We begin with this chapter on linear models.

3.2 LINEAR MODELS AND LINEAR EQUATIONS

Imagine a warehouse and loading dock with a load of copier paper to be transferred to a semitruck for shipment. The load consists of 100 pallets of paper that can be moved from the warehouse to the truck by a forklift. The first pallet has been loaded and contains 40 boxes and includes 4 boxes of colored paper. The remainder of the pallets contains 36 boxes of white paper each. The forklift can move one pallet at a time. Given the layout of the warehouse, it will take from one to three minutes for the forklift to

TABLE 3.1   Truck Loading Data

Trip	Time (minutes)	Boxes in the Truck	First Difference
0	0	40	40
1	2	76	36
2	4	112	36
3	6	148	36
...			

get to a pallet, traverse the distance to the truck, and drop the pallet into position on the truck. The average amount of time to load a pallet can be assumed to be two minutes. Table 3.1 represents the first several parts of the load represented over time where the first load takes the pallet of 40, and subsequent loads take the regular pallets.

The first pallet contains 40 boxes, leaving 3564 (99 pallets × 36 boxes) left to load. The second and all subsequent loads contain 36 boxes, for a difference in the remaining load of 36 fewer boxes each time. The pattern of change associated with the loading of the truck can be characterized by looking at the amount of change in the boxes to be loaded over time. After the initial pallet is loaded, this can be observed to be a constant making this relationship linear. The exact equation for this model is

$$b = 36k + 40 \tag{3.1}$$

where:
   b is the number of boxes on the truck
   k is the number of trips made by the forklift

The accumulated time for any particular trip will be 2k since it takes an average of two minutes for each trip. We could use this linear equation to create a simple, linear model of the loading process that would track the number of loaded boxes over time. We can also use the relationships to determine how long it will take to load the truck. How long will it take?

The general linear equation is written as follows:

$$Y = aX + b \tag{3.2}$$

In this equation, b is called the Y intercept or the value of Y when X is zero. If we graph a line, this will be the place where the line crosses the Y axis. The value of a represents the slope of the line or the ratio between the

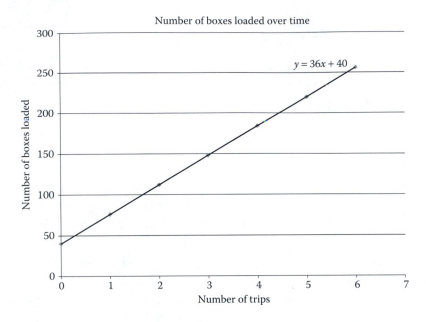

FIGURE 3.1   Linear equation graph showing boxes loaded over time.

change in Y versus the change in X. An illustration of Equation 3.1 for the loading of the paper pallets is shown as Figure 3.1.

> Any model where a change in the independent variable (the cause) causes a constant change in the dependent variable (the effect) is a linear model. A linear model can then be identified by calculating the slope or the ratio of change in Y versus change in X for each pair of data points. That slope will be constant if the relationship is linear.

There are a number of phenomena that can be represented by a linear model. One of these is a model of the deformation of a spring. The force related to the spring is measured in Newtons (N). One Newton is the force that will accelerate 1 kg of mass at a rate of 1 m/s².

The system is represented by Hooke's law:

$$F = kx \qquad (3.3)$$

where:
 F is the spring force (N)
 k is the spring constant (N/mm or N/cm)
 x is the deformation of the spring

When a force is applied to a spring, it returns to its original upstretched shape when released as long as the force exerted is within its elastic range. The spring constant is dependent upon the nature of the material used to make the spring as well as its winding and shape. The spring constant can be found experimentally by hanging dead weights on the spring. To see an illustration of this system, go to https://phet.colorado.edu/ and choose Play with Simulations. From there, choose enter spring in the search box and then choose the Hooke's Law (HTML5) simulation. Start the simulation and then choose the Intro section. Check all of the measurement boxes and then do your own experiment. For a given spring constant, apply forces of 10, 20, 30, and 40 N and record the values of the deformation. You can plot the force versus the displacement using graph paper or by entering the data into MATLAB® or Python and by creating a simple plot. You should observe that the relationship is linear.

The resistance of a material to an electric current is also represented as a linear equation:

$$R = \frac{\rho l}{a} \tag{3.4}$$

where:
   R is the resistance
   ρ is the resistivity
   l is the length of the sample
   a is the cross-sectional area of the sample

3.3 LINEAR INTERPOLATION

There are situations in which a sample of data is presented that provides selective pairs of points representing their relationship. One example is a reference table such as Table 3.2 that shows the relationship between altitude and the density of air. If we wanted to estimate the air density at 2250 m, we can use the values at 2000 and 2500 m to make a linear interpolation, assuming that the relationship is linear.

This assumption makes it possible to use the ratio between the known points and the target value to solve for the estimate of the unknown value. We can define the known values of the lookup column as (x_1, x_2) and the corresponding values of the related column as (y_1, y_2). We have the target

TABLE 3.2   Air Density Changes with Altitude

Altitude (m)	Air Density (kg/m³)
0	1.225
500	1.167
1,000	1.112
1,500	1.058
2,000	1.006
2,500	0.957
3,000	0.909
3,500	0.863
4,000	0.819
4,500	0.777
5,000	0.736
6,000	0.660
7,000	0.590
8,000	0.526
9,000	0.467
10,000	0.413
11,000	0.365
12,000	0.312
13,000	0.226
14,000	0.228
15,000	0.195

value of the lookup function x_0 but wish to estimate the value of y_0 by linear interpolation. Then:

$$\frac{y_0 - y_1}{x_0 - x_1} = \frac{y_2 - y_1}{x_2 - x_1} \tag{3.5}$$

This can be solved for y_0:

$$y_0 = y_1 + (x_0 - x_1)\left(\frac{y_2 - y_1}{x_2 - x_1}\right) \tag{3.6}$$

For our example in Table 3.2, this would be as follows:

$$y_0 = 1.006 + (2250 - 2000)\left(\frac{1.006 - 0.957}{2000 - 2500}\right) = 0.9815 \tag{3.7}$$

Linear interpolation might also be used when we have a sample of spatially distributed data points representing a continuous variable such as air temperature or precipitation. If we are willing to assume that the values of the unknown points between the samples are linearly distributed, we can use linear interpolation to estimate the intermediate values. However, there are often intervening variables that would violate that assumption so we must apply it with caution. For example, the air temperature is greatly impacted by the nature of the land cover. Surfaces such as asphalt and concrete absorb more solar radiation and reradiate it back into the atmosphere as heat, whereas vegetated areas like forests or grasslands will reflect more of the sunlight keeping those areas cooler. We would have to ascertain that the land cover is relatively constant between two sample points to make the case that linear interpolation of the data is justified.

3.4 SYSTEMS OF LINEAR EQUATIONS

There are a number of phenomena that are characterized by systems of linear equations where there are multiple linear equations. These are represented with a series of linear equations and multiple unknowns that must be solved simultaneously. Examples include a number of engineering problems including voltage in electrical circuits, chemical reactors, and static forces in structures. In the social sciences, econometric models of consumer and behavior, models of voting behavior, and models of land use and population change have been developed using simultaneous linear equations.

The solution of these problems requires an understanding of linear algebra and the algorithms used to solve the problems on a computer. Details of those operations are beyond the scope of this chapter. Both MATLAB and Python have built-in routines to solve these problems. In MATLAB, the matrix functions that will be introduced in a later chapter are used to solve simultaneous linear equations. In Python, the NumPy library has a set of routines that perform the same functions.

For those who are interested in some examples and further discussion of these methods, some example problems from engineering can be found in Chapra (2008, pp. 203–235) and Moaveni (2014, pp. 708–711). Landau et al. (2015, pp. 145–149) offer some examples in Python with exercises related to physics problems.

3.5 LIMITATIONS OF LINEAR MODELS

Linear models can offer a good approximation of certain phenomena but need to be used with special caution when making forecasts. All phenomena reach limits as you approach the extremes of their distributions—whether at the low end (zero or negative values for those that can go negative) or at the high end (at values that exceed normal behavior or result in nonlinear behavior).

Extrapolating beyond the available data used to create or validate a model is prone to major errors, whether the phenomenon we are modeling is linear or nonlinear in nature. For systems which we approximated with a linear representation, extrapolation may be doubly dangerous as there are very few circumstances where something will increase without reaching some limits that cause the system behavior to change.

For example, if we think about the truck loading example, we could forecast adding additional forklifts to speed the process. However, at some point, the average amount of time it takes for a forklift to complete a cycle of pickup and delivery will slow as additional forklifts will need to wait in line to get into or out of the truck or avoid each other inside the warehouse. If we did not think about this, we could easily add too many forklifts and forecast a directly proportional, linear reduction in loading times. Since forklifts and their operators are expensive, we could potentially raise costs without the requisite benefits of efficiency that we might desire.

EXERCISES

1. Examine the three datasets shown in Table 3.3. Which of these are linear relationships?

2. A new fast food chain has studied its success in its initial metropolitan market. They found that their net income from the opening and operation of each site could be represented by this linear equation:

$$I = b * 275,000 - 100,000 \tag{3.8}$$

TABLE 3.3 Example Datasets for Assessment of Linearity

Dataset 1		Dataset 2		Dataset 3	
X	Y	X	Y	X	Y
3	14	1	21	1.5	−0.5
6	20	2	35	2.5	4.5
9	26	3	57	3.5	9.5
12	32	4	87	4.5	14.5
15	38	5	125	5.5	19.5
18	44	6	171	6.5	24.5
		7	225	7.5	29.5
		8	287	8.5	34.5

where I represents the net revenue in dollars over one year, b represents the number of branches, and the constant represents the upfront costs of creating a new location. They have tested this relationship in their current market where they have 10 branches. If they are going to open a new market in a different metropolitan region, what would their net income be if they opened 4 branches, 8 branches, and 12 branches? Will this model work for all of those situations? What are the uncertainties to applying this model in this way?

3. We have implemented a model of the time it will take to get from our home to work comparing two major potential routes: one using local and major streets, and one that makes part of the trip on an Interstate highway. Use the materials on the book website to download the instructions and the partially completed code in either MATLAB or Python. Complete the code and make the model runs required to answer the questions about this model.

4. Water towers are used to store water and to release the treated water under sufficient pressure to reach the buildings in its service area and maintain pressure to keep the water from becoming contaminated. Using the materials on the book website to download the instructions and the partially completed code that predicts how tall the water tower needs to be to achieve the minimum necessary pressure. Complete the code in either MATLAB or Python and make the runs necessary to answer the questions about this model.

REFERENCES

Chapra, S. C. 2008. *Applied Numerical Methods with MALAB for Engineers and Scientists*, 2nd ed. Boston, MA: McGraw-Hill.

Landau, R. H., J. P. Manuel, and C. C. Bordeainu. 2015. *Computational Physics Problem Solving with Python*, 3rd ed. Weinheim, Germany: Wiley.

Moaveni, S. 2014. *Engineering Fundamentals*, 5th ed. Boston, MA: Cengage Learning.

CHAPTER 4

Array Mathematics in MATLAB® and Python

4.1 INTRODUCTION TO ARRAYS AND MATRICES

Thus far, our programming has focused on relatively simple representations of mathematical expressions and the syntax that can be used to declare and operate on variables. We have created arrays and matrices for the purpose of holding a range of values as input data or the intermediate or final values of our calculations. The outputs were calculated by applying the same formula to each element of the array and stored as values in the output arrays.

Beyond these simple calculations, both programming environments have the capability of executing matrix mathematics functions, which have several advantages. First, some calculations can be applied to the entire matrix at one time. In doing so, the code is simpler to write and the execution is much faster than the incremental calculation of individual elements contained in a loop. As problems become larger, the savings in computational resources can become very significant. We provide some examples in the exercises at the end of the chapter.

For a wide range of other calculations in science and engineering, matrix algebra is the most direct way of solving a variety of problems. Although linear and matrix algebra is not required to complete the modeling work presented in this book, it is still important to understand the nature of those calculations and their associated programming syntax so that the appropriate syntax can be used in the problems we present. Toward this end, we present a brief overview matrix mathematics. This is followed by the presentation of the programming syntax in MATLAB® and Python, respectively. For those with a background in matrix mathematics, we provide some additional examples and references at the end of the chapter.

4.2 BRIEF OVERVIEW OF MATRIX MATHEMATICS

Matrix notation represents a matrix in row and column order. Thus a 2×3 matrix as shown in Equation 4.1 has two rows and three columns. If we define it as the matrix [A], we can illustrate some matrix operations. If we want to multiply the matrix by a scalar (a single constant), we have to multiply each element by the scalar. Thus, [A]*2 will give an output matrix with the values 2, 4, 6, 8, 10, 12.

$$[A] = \begin{bmatrix} 1 & 2 & 3 \\ 4 & 5 & 6 \end{bmatrix} \tag{4.1}$$

Similarly, matrix addition and subtraction are also done element by element. However, for the calculations to work correctly, the matrices must have the same dimensions. Equations 4.2 through 4.4 illustrate the results of matrix addition and subtraction:

$$[B] = \begin{bmatrix} 10 & 11 & 12 \\ 13 & 14 & 15 \end{bmatrix} \tag{4.2}$$

$$[A] + [B] = \begin{bmatrix} 11 & 13 & 15 \\ 17 & 19 & 21 \end{bmatrix} \tag{4.3}$$

$$[B] - [A] = \begin{bmatrix} 9 & 9 & 9 \\ 9 & 9 & 9 \end{bmatrix} \tag{4.4}$$

Another operation we can perform on a matrix is to get its transpose. The transpose of a matrix makes the rows into columns and columns into rows. This transformation can be used to transform row vectors into column vectors or to provide a different view of a data table.

The calculations in matrix algebra for multiplication are not the same as the element by element procedure in addition and subtraction. A matrix of size m × n can only be multiplied by a matrix that is of size n × p.

That is the number of rows in the first matrix must equal the number of columns in the second matrix. The product is then a matrix of size m × p. Matrix multiplication is not commutative—that is the matrix multiplication of [A] × [B] will not produce the same result as [B] × [A].

The computation of the multiplication is the sum of the scalar products of *ith* row of the first matrix with the elements of the *jth* column of the second matrix. More specifically, the calculation:

$$[C] = [A] \times [B] \tag{4.5}$$

where:
    [A] is of the size m × n
    [B] is of the size n × p
    [C] is of the size m × p

Then

$$c_{ij} = \sum_{k=1}^{n} a_{ik} b_{kj} \tag{4.6}$$

where i, j, and k are the row and column indices for the input and output matrices.

This is illustrated by Equations 4.7 through 4.9:

$$[A] = \begin{bmatrix} 1 & 1 & 3 \\ 2 & 1 & 0 \end{bmatrix} \tag{4.7}$$

$$[B] = \begin{bmatrix} 1 & 2 \\ 0 & 2 \\ 2 & 1 \end{bmatrix} \tag{4.8}$$

$$[C] = \begin{bmatrix} 7 & 7 \\ 2 & 6 \end{bmatrix} \tag{4.9}$$

There are also several types of special matrices that are built in to both programming environments. One of these is the identity matrix. The identity matrix is a matrix that has ones in the diagonal and zeros everywhere else. The identity matrix is important in understanding the concept of the inverse of a matrix. When the inverse of a matrix is multiplied by

the original matrix, that product is the identity matrix. In matrix notation, the inverse of a matrix [A] is often represented as [A′]. Thus:

$$[A]\,[A'] = [I] \tag{4.10}$$

where I is the identity matrix. Only square matrices have an inverse.

There are several applications of these concepts but a full review is beyond
the scope of this section. However, for those with experience with linear and
matrix algebra, we provide an example exercise at the end of the chapter. The
remainder of the chapter reviews the matrix syntax of MATLAB and Python.

4.3 MATRIX OPERATIONS IN MATLAB®

In Chapter 2, we introduced several methods for creating a row vector.
The simplest of those is to type the vector values between square brackets,
separated by spaces, for example, [1 2 3]. You can create column vectors by
separating the elements by semicolons; for example, [1; 2; 3] will generate
the column vector

$$\begin{bmatrix} 1 \\ 2 \\ 3 \end{bmatrix}$$

To create a matrix, you can combine this syntax. [1 2 3; 4 5 6] will generate
the matrix

$$\begin{bmatrix} 1 & 2 & 3 \\ 4 & 5 & 6 \end{bmatrix}$$

(from Equation 4.1).

In Equation 4.3, we introduced the concept of matrix addition. If the
variables A and B contain the matrices described in Equations 4.1 and 4.2,
then to add the matrices A and B together, you can simply type

$$A + B$$

Matrix subtraction is similar; Equation 4.4 shows subtracting matrix A
from matrix B. The syntax is

$$B - A$$

Matrix multiplication is also straightforward. If A and B are the matrices in Equations 4.7 and 4.8, you can multiply them to get the matrix in Equation 4.9 using the following syntax:

$$C = A * B$$

To transpose a matrix—it must be square—you can use the transpose operator.

$$C'$$

4.4 MATRIX OPERATIONS IN PYTHON

If you will recall the instructions in Chapter 2 on creating arrays in Python, that we first need to import the NumPy library using *import numpy as np*, we can create row vectors with np.array([1,2,3]) (which will create the vector [1 2 3]). To create a column vector

$$\begin{bmatrix} 1 \\ 2 \\ 3 \end{bmatrix}$$

use the syntax np.array([[1], [2], [3]]). To create a matrix, such as the one in Equation 4.1, you can combine the syntax and use np.array([[1,2,3], [4,5,6]]) to create

$$\begin{bmatrix} 1 & 2 & 3 \\ 4 & 5 & 6 \end{bmatrix}$$

In Equation 4.3, we introduced the concept of matrix addition. If the variables A and B contain the matrices described in Equations 4.1 and 4.2, then to add the matrices A and B together, you can simply type

$$A + B$$

Matrix subtraction is similar; Equation 4.4 shows subtracting matrix A from matrix B. The syntax is

$$B - A$$

Matrix multiplication is also straightforward. If A and B are the matrices in Equations 4.7 and 4.8, you can multiply them to get the matrix in Equation 4.9 using the following syntax:

$$C = np.dot(A, B)$$

To transpose a matrix—it must be square—you can use the transpose function.

$$np.transpose(C)$$

EXERCISES

Given matrices

$$A = \begin{bmatrix} 1 & 2 & 3 \\ 4 & 5 & 6 \end{bmatrix}, B = \begin{bmatrix} 1 & 2 \\ 3 & 4 \end{bmatrix}, C = \begin{bmatrix} 1 & 2 \\ 3 & 4 \\ 5 & 6 \end{bmatrix}, \text{ and } D = \begin{bmatrix} 7 & 8 & 9 \\ 10 & 11 & 12 \end{bmatrix},$$

calculate the following:

1. $A - C'$

2. $C' + 3D$

3. BA

4. CB

5. B^4

6. AA'

7. $D'D$

8. For those with a background in linear algebra, download the file simultaneous.pdf from the book website and follow the instructions for MATLAB or Python solution of simultaneous linear equations.

CHAPTER 5

Plotting

5.1 PLOTTING IN MATLAB®

MATLAB® supports some sophisticated plotting tools, including 3D plotting and polar coordinates. In this chapter, we are going to focus on 2D plotting as a means to help visualize your models. Figure 5.1 shows an example plot generated in MATLAB to compare experimental results to the theoretical results for light intensity as a function of distance.

The basic 2D plot command is

```
plot(x, y)
```

where x and y are any vector names.

Both vectors must have the same number of elements. The plot command creates a single curve with the x values on the abscissa (horizontal axis) and the y values on the ordinate (vertical axis). The curve is made from segments of lines that connect the points that are defined by the x and y coordinates of the elements in the two vectors. The more elements there are to plot, the smoother the plot should look.

Given data in two vectors x and y, we can generate the plot in Figure 5.2 with the following code:

```
x=[1 2 3 5 7 7.5 8 10];
y=[2 6.5 7 7 5.5 4 6 8];
plot(x, y)
```

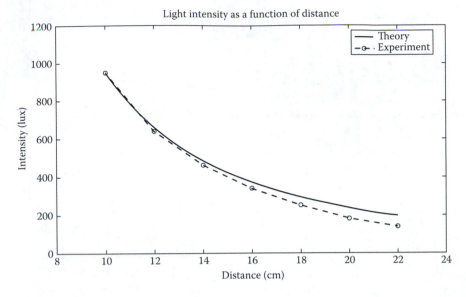

FIGURE 5.1 Example of a 2D plot in MATLAB.

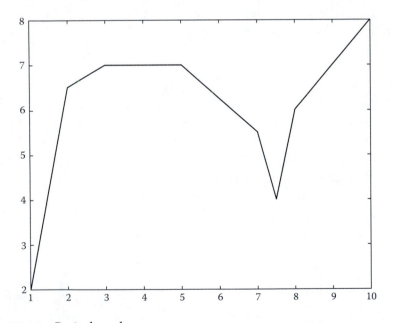

FIGURE 5.2 Basic data plot.

As soon as the plot command is executed, a new window will appear containing the plot. Since we have not given MATLAB any instructions about how to format the plot, it makes decisions about the axes, step sizes, line color, and so on. You can control all of these things.

Line specifiers—an optional parameter to the plot function—can be used to control the color of the line, the style of the line, and any data markers. The specifiers are entered as a coded string, but the specifiers can be in any order, and all of them are optional (i.e., 0, 1, 2, or all 3 can be included in a command).

Line Style	Specifier	Line Color	Specifier	Marker Type	Specifier
Solid	-	Red	r	X-mark	x
Dotted	:	Green	g	Circle	o
Dashed	--	Blue	b	Asterisk	*
Dash-dot	-.	Black	k	Point	.

For additional line markers and colors, please refer to MATLAB's built-in help by typing *help plot* in the Command Window, or using the *search documentation* box in the ribbon.

Let us plot the data in the following table to explore line specifiers.

Year	1988	1989	1990	1991	1992	1993	1994
Sales ($M)	127	130	136	145	158	178	211

To generate Figure 5.3, use the following code:

```
year=[1988:1:1994];
sales=[127 130 136 145 158 178 211];
plot(year, sales, '--r*')
```

We can also plot functions by using MATLAB to calculate the value of an equation for a specified set of input points. Consider Equation 5.1:

$$y = 3.5^{-0.5x} \cos(6x) \tag{5.1}$$

We can plot this equation over a range of −2 to 4 inclusive by first generating a vector with the endpoints and spacing we want, and then evaluating the equation for each point in the vector.

```
x=[-2: 0.01: 4];
y=3.5 .^ (-0.5*x) .* cos(6*x);
plot(x, y)
```

The results can be seen in Figure 5.4.

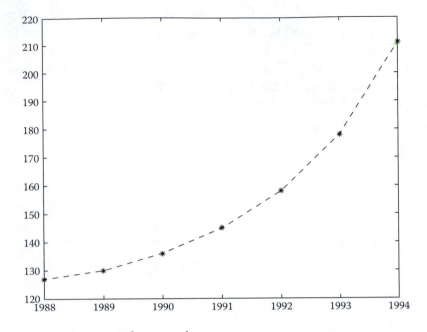

FIGURE 5.3   Line specifier example.

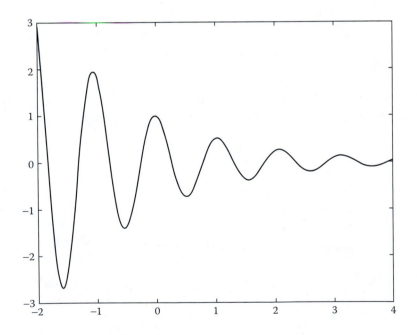

FIGURE 5.4   Plotting a function.

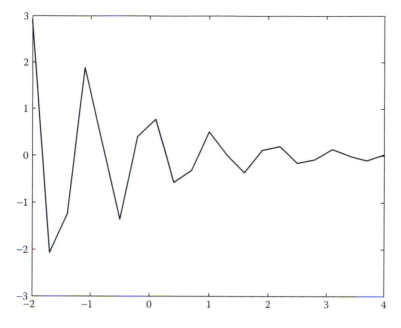

FIGURE 5.5   Low sampling resolution.

Plotting functions this way can be very sensitive to the number of points in the "x" vector. For example, if we change the distance between points on the range of −2 to 4 from 0.01 to 0.3, the plot looks like Figure 5.5. You should take care to ensure the plot you get accurately represents the function you wish to represent.

If you recall Figure 5.1, you should have noticed that we can put multiple graphs on the same plot. The first mechanism for doing so is providing multiple sets of data points to a single plot command. For example, to plot y versus x, v versus u, and h versus t on the same plot, you can issue this plot command:

```
plot(x, y, u, v, t, h)
```

By default, MATLAB will create each curve with a different color. You may also control the lines by providing line specifiers to each pair. For example:

```
plot(x, y,'-b', u, v,'−r', t, h,'g:')
```

To demonstrate this, let us plot a function, its first and second derivatives (as seen in Equations 5.2, 5.3, and 5.4), from −2 to 4 inclusive:

$$y = 3x^3 - 26x + 10 \tag{5.2}$$

$$y = 9x^2 - 26 \tag{5.3}$$

$$y = 18x \tag{5.4}$$

The code to generate the plot in Figure 5.6 is as follows:

```
x=[-2:0.01:4];
y=3*x.^3 - 26*x+6;
yd=9*x.^2 - 26;
ydd=18*x;
plot(x, y,'-b', x, yd,'--r', x, ydd,':k')
```

An alternative method for drawing multiple curves on a single plot is the *hold* command. You can issue a *hold on* to instruct MATLAB to keep the current plot and all axis properties, and to draw all subsequent plot commands in the existing plot window. Issuing a *hold off* command will return

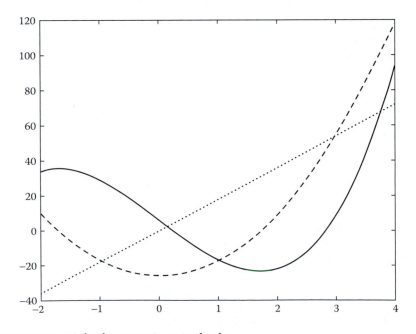

FIGURE 5.6   Multiple curves in a single plot.

MATLAB to the default mode where plot commands erase the previous plots and reset all axis properties before drawing new plots.

Let us use this technique to start constructing a publication-ready plot as demonstrated in Figure 5.1.

```
x=[10:0.1:22];
y=95000./x.^2;
x_data=[10:2:22];
y_data=[950 640 460 340 250 180 140];
plot(x, y,'-')
hold on
plot(x_data, y_data,'ro--')
hold off
```

This code generates Figure 5.7. A good start but a truly readable plot requires us to add axis labels, a title, legend, and potentially manipulate the axis limits. Let us learn how to do that now.

Once you have a plot, MATLAB provides multiple mechanisms for manipulating various plot properties. We are going to focus on using script-able commands (typed in the Command Window) as knowing those allows

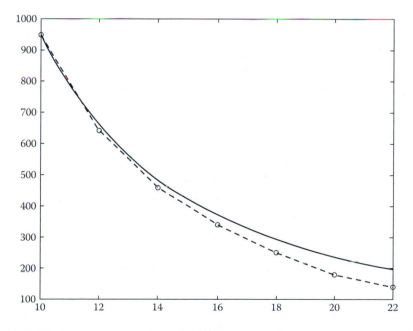

FIGURE 5.7   Data plot generated with *hold* command.

you to write programs that can generate complete plots; however, you can do everything we mention interactively via the menus in the GUI if you prefer.

To set a title, use the *title* function. It takes a single string as an argument. *xlabel* and *ylabel* functions also take a single string as an argument, and set the labels on the x and y axes, respectively. The *axis* function allows you to set the minimum and maximum limits of the x-axis and y-axis by providing a four element vector of the format [xmin   xmax   ymin   ymax]. To add a legend, you use the *legend* function, providing a string argument with the name of each curve, in the order the curves were added to the plot. Executing the following code will turn Figure 5.7 into Figure 5.1:

```
xlabel('DISTANCE (cm)')
ylabel('INTENSITY (lux)')
title('Light Intensity as a Function of Distance')
axis([8 24 0 1200])
legend('Theory','Experiment')
```

5.2 PLOTTING IN PYTHON

There is a Python library available called matplotlib that provides 2D-plotting tools and functions very similar to what MATLAB provides. It is included in the Anaconda installation we recommend, and it can be loaded into Python using the import command. We will be importing the library as *plt* to be consistent with the available documentation for matplotlib (which we recommend you to use as a reference, as the capabilities are extensive and are well beyond the basic introduction to plotting we provide).

```
import matplotlib.pyplot as plt
```

We also recommend changing your IPython preferences to instruct it to *not* use inline graphics. Our instructions are written assuming that you have set the IPython graphics backend setting to *Automatic*, as shown in Figure 5.8.

Figure 5.9 shows an example plot generated in Python to compare experimental results to the theoretical results for light intensity as a function of distance.

The basic 2D plot command is

```
plt.plot(x, y)
```

where x and y are any vector names.

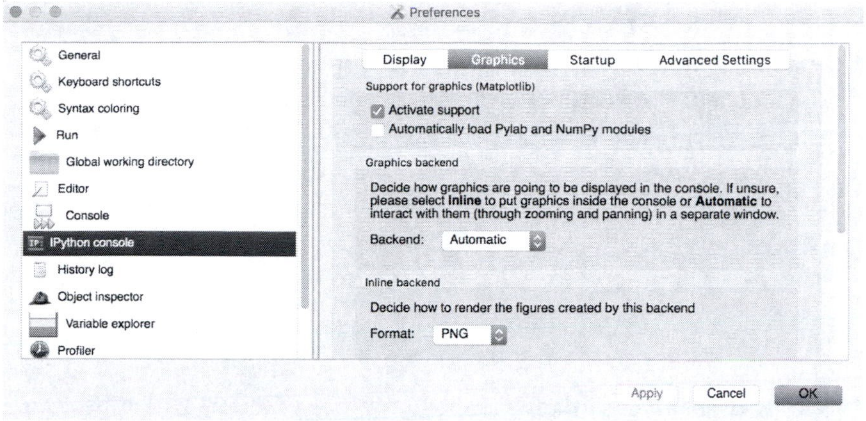

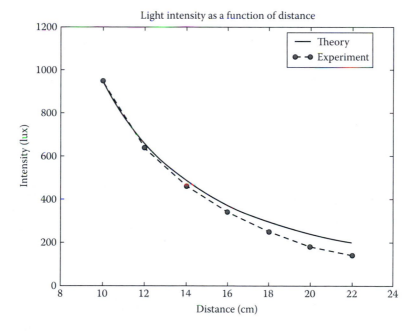

FIGURE 5.8   IPython graphics backend setting.

FIGURE 5.9   Example of a 2D plot in Python.

Both vectors must have the same number of elements. The plot command creates a single curve with the x values on the abscissa (horizontal axis) and the y values on the ordinate (vertical axis). The curve is made from segments of lines that connect the points that are defined by the x and y coordinates of the elements in the two vectors. The more elements there are to plot, the smoother the plot should look.

Given data in two vectors x and y, we can generate the plot in Figure 5.10 with the following code:

```
x=np.array([1,2,3,5,7,7.5,8,10])
y=np.array([2,6.5,7,7,5.5,4,6,8])
plt.plot(x, y)
```

As soon as the plot command is executed, a new window will appear containing the plot. Since we have not given Python any instructions about how to format the plot, it makes decisions about the axes, step sizes, line color, and so on. You can control all of these things.

Line specifiers—an optional parameter to the plot function—can be used to control the color of the line, the style of the line, and any data markers. The specifiers are entered as a coded string, but the specifiers can be in any order, and all of them are optional (i.e., 0, 1, 2, or all 3 can be included in a command).

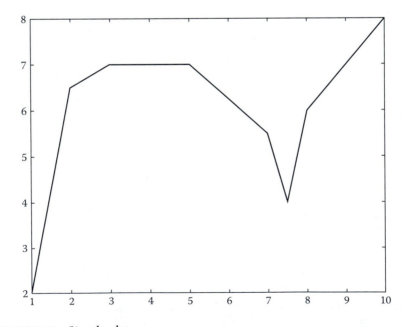

FIGURE 5.10 Simple plot.

Line Style	Specifier	Line Color	Specifier	Marker Type	Specifier
Solid	-	Red	r	X-mark	x
Dotted	:	Green	g	Circle	o
Dashed	--	Blue	b	Asterisk	*
Dash-dot	-.	Black	k	Point	.

For additional line markers and colors, please refer to the online documentation for matplotlib.

Let us plot the data in the following table to explore line specifiers:

Year	1988	1989	1990	1991	1992	1993	1994
Sales ($M)	127	130	136	145	158	178	211

To generate Figure 5.11, use the following code:

```
year=np.arange(1988,1994.1,1)
sales=np.array([127,130,136,145,158,178,211])
plot(year, sales, '--r*')
```

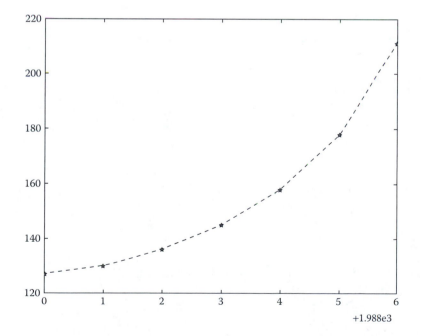

+1.988e3

FIGURE 5.11   Line specifier example.

We can also plot functions by using Python to calculate the value of an equation for a specified set of input points. Consider Equation 5.5:

$$y = 3.5^{-0.5x} \cos(6x) \tag{5.5}$$

We can plot this equation over a range of −2 to 4 inclusive by first generating a vector with the endpoints and spacing we want, and then evaluating the equation for each point in the vector.

```
x=np.arange(-2,4.001,.01)
y=3.5**(-0.5*x)*np.cos(6*x)
plt.plot(x, y)
```

The results can be seen in Figure 5.12.

Plotting functions this way can be very sensitive to the number of points in the "x" vector. For example, if we change the distance between points on the range of −2 to 4 from 0.01 to 0.3, the plot looks like Figure 5.13. You should take care to ensure the plot you get accurately represents the function you wish to represent.

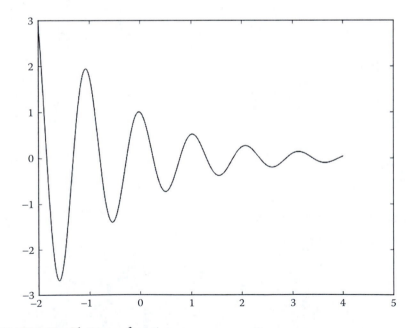

FIGURE 5.12   Plotting a function.

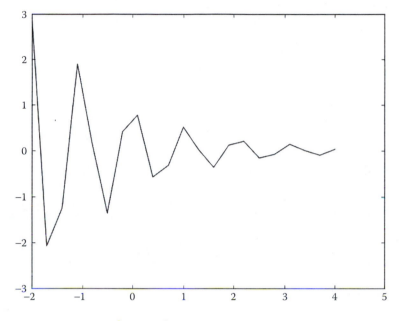

FIGURE 5.13   Low sampling resolution.

If you recall Figure 5.9, you should have noticed that we can put multiple graphs on the same plot. The first mechanism for doing so is providing multiple sets of data points to a single plot command. For example, to plot y versus x, v versus u, and h versus t on the same plot, you can issue this plot command:

```
plt.plot(x, y, u, v, t, h)
```

By default, matplotlib will create each curve with a different color. You may also control the lines by providing line specifiers to each pair. For example:

```
plot(x, y,'-b', u, v,'-r', t, h,'g:')
```

To demonstrate this, let us plot a function, its first and second derivatives (as seen in Equations 5.6 through 5.8), from −2 to 4 inclusive:

$$y = 3x^3 - 26x + 10 \tag{5.6}$$

$$y = 9x^2 - 26 \tag{5.7}$$

$$y = 18x \tag{5.8}$$

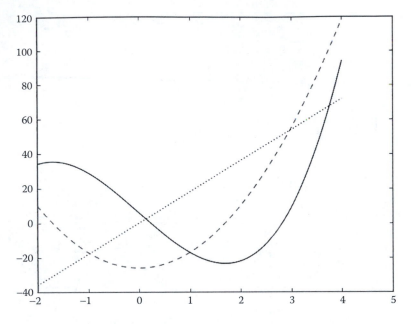

FIGURE 5.14 Multiple curves in a single plot.

The code to generate the plot in Figure 5.14 is as follows:

```
x=np.arange(-2,4.001,.01)
y=3*x**3 - 26*x+6
yd=9*x**2 - 26
ydd=18*x
plt.plot(x, y,'-b', x, yd,'--r', x, ydd,':k')
```

In matplotlib, additional plot commands will be automatically added to the existing plot figure. You can work on multiple figures by calling the *figure* function. For example, to open a second figure for subsequent *plot* function calls to draw in, you can type "figure(2)". You can switch between plot windows for future commands by passing the figure function the number of the figure window you wish to make active.

Let us start constructing a publication-ready plot as demonstrated in Figure 5.9. We are going to add a label to our plot commands now, which will be used to generate a legend later.

```
x=np.arange(10,22.01,0.1)
y=95000/x**2;
x_data=np.arange(10,22.01,2)
y_data=np.array([950,640,460,340,250,180,140])
plt.plot(x, y,'-', label='Theory')
plt.plot(x_data, y_data,'ro--', label='Experiment')
```

This code generates Figure 5.15. A good start but a truly readable plot requires us to add axis labels, a title, legend, and potentially manipulate the axis limits. Let us learn how to do that now.

To set a title, use the *title* function. It takes a single string as an argument. *xlabel* and *ylabel* functions also take a single string as an argument, and set the labels on the x and y axes, respectively. The *axis* function allows you to set the minimum and maximum limits of the x-axis and y-axis by providing a four-element vector of the format [xmin   xmax   ymin   ymax].

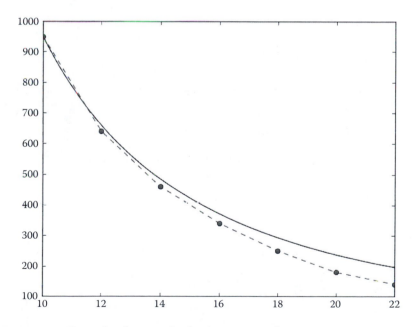

FIGURE 5.15   Data plot from multiple plot commands.

To add a legend, you use the *legend* function, which will use the labels you provided when generating the plot(s) earlier. Executing the following code will turn Figure 5.15 into Figure 5.9:

```
plt.xlabel('DISTANCE (cm)')
plt.ylabel('INTENSITY (lux)')
plt.title('Light Intensity as a Function of Distance')
plt.axis([8,24,0,1200])
plt.legend()
```

EXERCISES

1. Make two separate plots of the function

$$f(x) = 0.01x^4 - 0.45x^2 + 0.5x - 2$$

one for $-4 \leq x \leq 4$ and one for $-8 \leq x \leq 8$.

2. Plot the function

$$f(x) = \frac{x^2 - 4x - 5}{x - 2}$$

for $-4 \leq x \leq 8$. Notice that the function has a vertical asymptote at $x = 2$. Plot the function by creating two vectors for the domain of x, the first vector with elements from -4 to 1.7, and the second vector with elements from 2.3 to 8. For each x vector, create a y vector with the corresponding values of y according to the function. To plot the function, make two curves in the same plot.

3. You have an electrical circuit that includes a voltage source v_s with an internal resistance r_s and a load resistance R_L. The power P dissipated in the load is given by

$$P = \frac{v_s^2 R_L}{(R_L + r_s)^2}$$

Plot the power P as a function of R_L for $1\Omega \leq R_L \leq 10\Omega$ given that $v_s = 12V$ and $r_s = 2.5\Omega$.

4. The Gateway Arch in St. Louis is shaped according to the following equation:

$$y = 693.8 - 68.8 \, \cosh\left(\frac{x}{99.7}\right)$$

Make a plot of the arch. Draw a horizontal line at ground level ($y = 0$), $x = \pm 99.7$ acosh ($693.8/68.8$).

5. In astronomy, the relationship between the relative temperature T/T_{SUN} (temperature relative to the sun), relative luminosity L/L_{SUN}, and relative radius R/R_{SUN} a star is modeled by

$$\frac{L}{L_{SUN}} = \left(\frac{R}{R_{SUN}}\right)^2 \left(\frac{T}{T_{SUN}}\right)^4$$

The Hertzsprung–Russell (HR) diagram is a plot of L/L_{SUN} versus the temperature. The following data are given:

	Sun	Spica	Regulus	Alioth	Barnard's Star	Epsilon Indi	Beta Crucis
Temperature (K)	5840	22,400	13,260	9,400	3,130	4,280	28,200
L/L_{SUN}		13,400	150	108	0	0.15	43,000
R/R_{SUN}		7.8	3.5	3.7	0.18	0.76	8

To compare the data with the model, plot a HR diagram. The diagram should have two sets of points. One uses the values of L/L_{SUN} from the table (use asterisk markers), and the other uses values of L/L_{SUN} that are calculated by the equation using R/R_{SUN} from the table (use circle markers). In the HR diagram both axes are logarithmic. In addition, the values of the temperature on the horizontal axis are decreasing from left to right. Label the axes and use a legend.

6. The position x as a function of time that a particle moves along a straight line is given by

$$x(t) = -0.1t^4 + 0.8t^3 + 10t - 70$$

The velocity v(t) is determined by the derivative of x(t) with respect to t, and the acceleration a(t) is determined by the derivative of v(t) with respect to t.

Derive the expressions for the velocity and acceleration of the particle, and make plots for the position, velocity, and acceleration as a function of time for $0 \le t \le 8$. Time t is measured in seconds, and position x is measured in meters. Make three plots, one for position, one for velocity, and one for acceleration. Label the axes appropriately with correct units.

7. A resistor, $R = 2\Omega$, and an inductor, $L = 1.7H$, are connected to a voltage source in series (RL circuit). When the voltage source applies a rectangular voltage pulse with an amplitude of $V = 24V$ and a duration of 0.5 s, the current i(t) in the circuit as a function of time is given as follows:

$$i(t) = \left(\frac{V}{R}\right)\left(1 - e^{\frac{-Rt}{L}}\right) \quad \text{for} \quad 0 \le t \le 0.5\,s$$

$$i(t) = e^{-\frac{Rt}{L}}\left(\frac{V}{R}\right)\left(e^{\frac{0.5R}{L}}\right) \quad \text{for} \quad 0.5 \le t\,s$$

Make a plot of the current as a function of time for $0 \le t \le 5$ s.

CHAPTER 6

Problem Solving

Problem solving includes experience, knowledge, process, and art. In this chapter, we want to focus on computer-assisted problem solving; or programming. Computers are good tools for solving rule-based problems. Most problems have more than one solution. Thus the programmer must tie in his or her prior experience, knowledge, and understanding of the problem to produce a solution that most efficiently solves the problem. Problem solving is an art, in that the problem solver comes up with his or her own unique solution to the problem.

6.1 OVERVIEW

Problem solving requires a combination of science and art. On the technical side, we have math, chemistry, physics, mechanics, and so on, whereas on the artistic side, we have things such as judgment, prior experience, common sense, know-how, and so on.

We will use what we will call the *engineering method* to approach solving problems. The steps, broadly, consist of the following:

- Recognize and understand the problem

- Gather data (and verify its accuracy)

- Select guiding theories and principles

- Make valid, safe assumptions when necessary

- Solve the problem

- Verify the results

- Present the solution

If time permits and the results are not what you expected, you may wish to go back a few steps to improve the solution. What are the steps in detail?

Identify the problem: You should be able to create a clear written statement of the problem to be solved. If you cannot do this, you will struggle to come up with a good solution.

Determine what is required for the solution: What is known? What is unknown? Are there any restrictions or limitations? Are there any special cases?

Develop a step-by-step plan (algorithm): How are we going to solve the problem? What steps does our program need to take?

Outline the solution in a logic diagram: It is usually helpful to outline the solution to a programming problem in a logic diagram. This step will likely create something that could be translated into a high-level version of the program used to solve the problem.

Execute the plan: Write the code. Keep track of what works, and what does not (source control, such as git, is often helpful but outside the scope of this book). It can be helpful to write and test small portions of the solution independently to make sure that those parts work as expected. This will allow you to build on smaller successes and to *eat the elephant* one bite at a time.

Analyze the solution: Revise the plan and reexecute as needed. Keep the good parts of the plan, and discard the not so good parts. We talk in greater detail about verifying and validating models in Chapter 12.

Report/document the results: Let your team know how your idea worked in writing. Good documentation and records of your development and analysis will be helpful at this stage.

6.2 BOTTLE FILLING EXAMPLE

A specific example will be useful for making this more concrete. The problem is simple: fill a bottle with stones. We need to document any assumptions that may be required, and then write a step-by-step procedure for solving the problem. Remember when doing these things, you need to *think like a machine*. We want to avoid assumptions and steps in our procedure that are too broad and vague.

First, what assumptions do we need to make? This is one possible list, but there could be other assumptions as well.

- Bottle is present

- Stones are present

- There are enough stones to fill the bottle

- Bottle is empty (or at least not full)

- Some (or all) stones fit through the opening

What might an algorithm look like? Remember, this is partially an art, and our algorithm here is not the only possible solution. There is not necessarily a *right* answer, only answers that work, and ones that do not.

1. Set bottle upright near stones.

2. Pick up a stone and try to put it in the bottle.

3. If the stone is too large to fit, discard the stone, and go to step #2.

4. Otherwise, the stone fits, so put it in the bottle.

5. Check to see if the bottle is full. If not, then go to step #2.

6. The bottle is full, so stop.

Now we have an algorithm for solving the bottle filling problem. We will return to this example later in the chapter.

6.3 TOOLS FOR PROGRAM DEVELOPMENT

A variety of tools and techniques can be used in the process of program development and are useful for organizing the tasks in problem solving. Many of these are focused on the development or formulation of algorithms, the representation of algorithms, and the refinement or structuring of algorithms. There are three techniques that we will discuss: top–down design, pseudocode, and logic diagrams or flowcharts. You will not necessarily use any single technique, but likely a blend of the techniques depending on the problems you are attempting to solve and your comfort with the techniques.

6.3.1 Pseudocode

Pseudocode is an artificial and informal *programming* language that helps programmers to develop algorithms. Unlike MATLAB® or Python, there is no standard grammar or syntax, and no interpreter to execute code. It is simply writing steps in an algorithm in a code-like manner, rather than using something like full English sentences. We will use examples later in this chapter.

6.3.2 Top–Down Design

With top–down design, you will begin with a single statement that conveys the overall function of the program. It is a complete (but simple) representation of the program. Then, you divide that top statement into a series of smaller tasks and list them in the order in which they must be performed. Next, you will refine each of these smaller tasks into yet smaller steps, defining specific internal and external data required. Continue this refinement process until all tasks are broken into small tasks that can be simply programmed as computer code. At this point, the algorithm is complete. When combining this technique with pseudocode, writing the actual program is normally straightforward.

Let us examine an example. First, the problem:

> Twenty students in a class have taken their first quiz. The grades (in the range of 0 to 100) are available. Determine the class average on the quiz.

What items are known? The grades, the possible range of legitimate grades, and that the user of the program does not know MATLAB or Python are all pieces of information we know. What is unknown? How the grades are accessed by the program? So, we will list an assumption that the grades will be input one by one, with a flag value (−99) to indicate when done.

The first step is to write the top-level statement of the program using pseudocode.

1.0 Determine the class average for the quiz

While this statement accurately captures the program's purpose, it does not provide much insight into how to write the code. So the next step is to refine this statement into more detailed statements.

1.0 Determine the class average for the quiz

    1.1 Initialize variables

    1.2 Input and sum the quiz grades

    1.3 Calculate and print the class average

Progress! Now, we need to further refine these (and use pseudocode to describe these steps, where we can).

1.0 Determine the class average for the quiz

    1.1 Initialize variables

        1.1.1 Initialize a running total to zero

        1.1.2 Initialize grade count to zero

    1.2 Input and sum the quiz grades

        1.2.1 Request and get the first grade

        1.2.2 Add this grade to the running total

        1.2.3 Add one to grade count

        1.2.4 Input next grade

        1.2.5 Add this grade into the running total

        1.2.6 Add one to grade count

        1.2.7 If grade input is −99, go to task 1.3, otherwise repeat steps from 1.2.4

    1.3 Calculate and print the class average

        1.3.1 Set the class average to the running total divided by the grade count

        1.3.2 Print the average

6.3.3 Flowcharts

Flowcharts—diagrams that describe the logic flow of the program—are very powerful tools in program development. They are not a complete description of the program, and not the only tool to use, but can be very

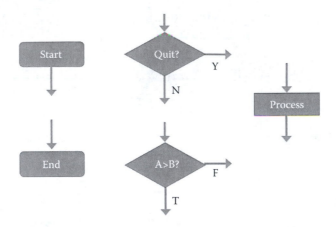

FIGURE 6.1   Flowchart symbols.

helpful for envisioning the various paths through the program required to solve a problem. There are no formal standards for flowcharts, but there are some common guidelines we will lay out.

There are a number of standard shapes used in flowcharts, and sticking to this convention will make it easier for others to understand your flowcharts. For example, the beginning and ending of a program are normally shown within a rectangle with rounded corners, containing text such as *start*, *begin*, and *end*, or in the case of a subprogram, *enter* for example. Arrows are used to designate the flow of the program, and normally the arrow will point to the next sequential program block in the flowchart. Any time a decision has to be made by the program, a selection structure is used. This is a diamond shape, with one entrance and two exits (labeled with the choice—yes or no, true or false, etc.) and containing a question (such as *Quit?* or *A < B?*). General processing blocks are commonly represented by a rectangle. These commands could be completing some mathematical equation for example. See Figure 6.1 for examples.

6.4 BOTTLE FILLING EXAMPLE CONTINUED

We now know enough to generate a full flowchart for the bottle filling example in Section 6.2. Figure 6.2 contains a flowchart for *one possible* solution to the bottle filling problem. If you compare the flowchart to the

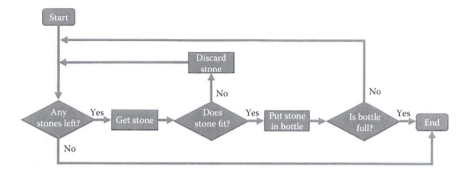

FIGURE 6.2   Bottle filling example flowchart.

top–down example created in Section 6.2, you will note that the algorithms are different.

EXERCISES

1. Using the top–down method, please describe an algorithm for a program to solve the real roots of a quadratic equation $ax^2 + bx + c = 0$. Your algorithm should ask the user for the coefficients. To calculate the roots of the equation, your algorithm should calculate the discriminant D given by

$$D = b^2 - 4ac$$

If $D > 0$, the algorithm displays a message: *The equation has two roots* and then displays the roots.

If $D = 0$, the algorithm displays a message: *The equation has one root*, and then displays the root.

If $D < 0$, the algorithm displays a message: *The equation has no real roots*.

Create a flowchart for your algorithm.

2. Using the top–down method, please describe an algorithm for a program that calculates the cost of a car rental according to the following price schedule:

| | Rental Period | | |
Type of Car	1–6 days	7–27 days	28–60 days
Class B	$27 per day	$162 for 7 days, +$25 for each additional day	$662 for 28 days, +$23 for each additional day
Class C	$34 per day	$204 for 7 days, +$31 for each additional day	$810 for 28 days, +$28 for each additional day
Class D	Class D cannot be rented for less than 7 days	$276 for 7 days, +$43 for each additional day	$1136 for 28 days, +$38 for each additional day

The algorithm asks the user to enter the rental period and type of car. The algorithm should display the appropriate cost. If a period longer than 60 days is entered, a message "Rental is not available for more than 60 days" should be displayed. If a rental period of less than 6 days is entered for Class D, a message "Class D cars cannot be rented for less than 6 days" should be displayed. Create a flowchart for your algorithm.

CHAPTER 7

Conditional Statements

I N CODE WE HAVE written and executed so far, all of the commands have been executed in sequential order. More powerful programs and models will require the ability for the code to make decisions about which parts of the code should be executed, and about some of the mechanisms to control, which are presented in this chapter. Specifically, we are going to teach you how to select one of several possible program paths—distinct blocks of code to execute—based on a *condition*; an evaluation of a variable against some test state.

7.1 RELATIONAL OPERATORS

In order to compare a variable against a static condition (or against another variable), we need to introduce relational (or comparison) operators. Relational operators compare the operands (the values before and after the operator) and based on the operator, evaluate the comparison and tell us if the condition is true or false. Assume variable a holds 5 and variable b holds 10, then

Operator		Description	Example
<		If the value of the left operand is less than the value of the right operand, then the condition becomes true.	(a < b) is true.
<=		If the value of the left operand is less than or equal to the value of the right operand, then the condition becomes true.	(a <= b) is true.
>		If the value of the left operand is greater than the value of the right operand, then the condition becomes true.	(a > b) is not true.
>=		If the value of the left operand is greater than or equal to the value of the right operand, then the condition becomes true.	(a >= b) is not true.
==		If the values of two operands are equal, then the condition becomes true.	(a == b) is not true.
~=	MATLAB	If the values of two operands are not equal, then the condition becomes true.	(a != b) and (a ~= b)
!=	Python		are true.

Note: In MATLAB answers to these comparisons are returned as scalar values, 1 representing true and 0 representing not true.

7.2 LOGICAL OPERATORS

We can also do logical tests, comparing two operands to make decisions based on if those operands are true or not true. Combined with comparison operators, we can build fairly complicated conditional tests. If a is true, and if b is true, then

Python Operator	MATLAB Operator	Description	Example
and	&	If both the operands are true, then the condition becomes true.	(a and b) is true
or	|	If any of the two operands are true, then the condition becomes true.	(a or b) is true
not	~	Used to reverse the logical state of its operand.	not (a and b) is false

Note: In MATLAB, the operands are numeric, as is the output of the logical test. For example, any nonzero number is evaluated as *true*, and the result of a logical test that is true is the value 1. Not true logical tests return 0.

7.3 CONDITIONAL STATEMENTS

Conditional statements allow our programs to make decisions, and it is not too dissimilar to the way we (humans) make decisions. The way it works is that a condition is stated; if that condition is met, one set of actions is taken. If the condition is not met, either nothing is done, or a second set of actions is taken. For example,

> If I win the lottery,
>
>> I will quit college, buy a new car, and go fishing.
>
> If I do not win the lottery,
>
>> I will study harder so that I can get a better job.

7.3.1 MATLAB®

There are three forms of *if* statements that can be formed in MATLAB®. The most basic is *if-end*. In this form, a command group will be executed if the conditional statement is true, and not executed if the conditional statement is not true. Normal sequential code execution resumes after the *end* keyword.

```
if conditional statement
        command group
end
```

The second form is the *if-else-end* format. In this version, command group 1 is executed if the conditional statement is true, whereas command group 2 is executed if the conditional statement is not true.

```
if conditional statement
        command group 1
else
        command group 2
end
```

The final form is *if-elseif-else-end*. Here, we can chain multiple *if* statements together that are only evaluated if (all) previous if statements evaluated as not true. You can have any number of elseif statements, and the final *else* statement and command group is entirely optional. As soon as any command group is executed, the program will jump ahead to the *end* keyword.

```
if conditional statement 1
        command group 1
elseif conditional statement 2
        command group 2
...
elseif conditional statement #
        command group #
...
else
        command group n+1
end
```

This final form can be complicated, so a flowchart demonstrating how a *if-elseif-else-end* block with three possible command groups works is shown in Figure 7.1.

Let us use a more concrete example. Assume we wanted a program that would calculate a tip based on the size of the bill such that the minimum tip was $1.80, the program would tip 18% if the bill was less than $60, and it would tip 20% if the bill was greater than or equal to $60. If we convert that to a flowchart, we can see how it would look in Figure 7.2.

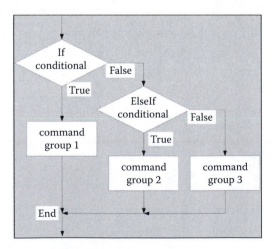

FIGURE 7.1   If-elseif-else-end flowchart.

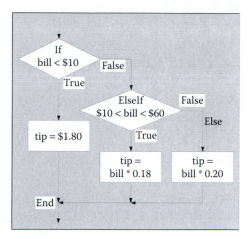

FIGURE 7.2   Tipping flowchart (MATLAB Syntax).

Here is the MATLAB code to implement that flowchart. Please note we are changing the display number format to *bank* to ensure the variables containing money display output at the right number of decimal points.

```
format bank
bill=input('Enter the amount of the bill (in dollars): ');
if (bill <= 10)
        tip = 1.8;
elseif (bill > 10) & (bill <= 60)
        tip = bill*0.18;
else
        tip = bill*0.2;
end
disp('The tip is (in dollars):')
disp(tip)
```

Remember the *end* keyword is required for every *if* command. There is no limit to the number of *if* commands used in a program, and there are many different ways to combine the *if-end*, *if-else-end*, and *if-elseif-else-end* formats to perform the same task. *Else* conditions are optional, always, and do not have conditional statements attached to them.

MATLAB also supports *switch-case* structures, which are similar to *if-elseif-else-end* statements. In a *switch-case* structure, we take a single scalar or string, and compare that to a number of different *cases* to pick a match. For example (we will explain the *while* statement in a future chapter), this code will display the value in x if that value is 100, 200, 300, or 400, and ask for the user's input again, and it will exit on any other value of x.

```
n = 1;
while (n ~= 0)
        x = input ('Resistance: ');
        switch x
                case 100
                disp (x)
                case 200
                disp (x)
                case 300
                disp (x)
                case 400
                disp (x)
                otherwise
                n = 0;
        end
end
```

7.3.2 Python

Python provides three forms of *if* conditional statements. The most basic is *if*. In this form, a command group will be executed if the conditional statement is true and not executed if the conditional statement is not true. Normal sequential code execution resumes after the indentation returns to the level of the if statement. (Python uses indentation to designate code blocks, and is thus very sensitive to white space.)

```
if conditional statement:
        command group
```

The second form is the *if-else* format. In this version, command group 1 is executed if the conditional statement is true, whereas command group 2 is executed if the conditional statement is not true.

```
if conditional statement:
        command group 1
else:
        command group 2
```

The final form is *if-elif-else*. Here, we can chain multiple *if* statements together that are only evaluated if (all) previous if statements evaluated as not true. You can have any number of elif statements, and the final *else* statement and command group is entirely optional. As soon as any command group is executed, the program will jump ahead past the rest of the command groups.

```
if conditional statement 1:
        command group 1
elif conditional statement 2:
        command group 2
. . .
elif conditional statement #:
        command group #
. . .
else:
        command group n+1
```

This final form can be complicated, so a flowchart demonstrating how a *if-elif-else* block with three possible command groups works is shown in Figure 7.3.

Let us use a more concrete example. Assume we wanted a program that would calculate a tip based on the size of the bill such that the minimum

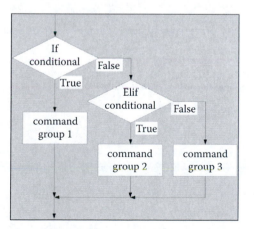

FIGURE 7.3   If-elif-else flowchart.

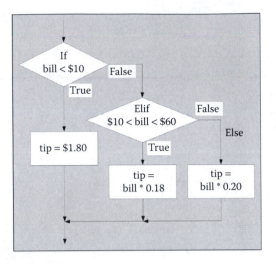

FIGURE 7.4   Tipping flowchart (Python Syntax).

tip was $1.80, the program would tip 18% if the bill was less than $60, and it would tip 20% if the bill was greater than or equal to $60. If we convert that to a flowchart, we can see how it would look in Figure 7.4.

Here is the Python code to implement that flowchart. Note that we must cast the input, which will be a string, to an integer type, and that we are applying some formatting to the output to only display two decimal places.

```
bill=int(input('Enter the amount of the bill (in dollars): '))
if (bill <= 10):
    tip=1.8
elif (bill>10)and(bill<=60):
    tip=bill*0.18
else:
        tip=bill*0.2
print('The tip is (in dollars): %0.2f'% (tip))
```

There is no limit to the number of *if* commands used in a program, and there are many different ways to combine the *if*, *if-else*, and *if-elif-else* formats to perform the same task. *Else* conditions are optional, always, and do not have conditional statements attached to them.

EXERCISES

1. Expanding on Problem 1 from Chapter 6, implement your algorithm. The problem description is restated as follows:

   Write a script called *quadroots* to solve the real roots of a quadratic equation $ax^2 + bx + c = 0$. Your algorithm should ask the user for the coefficients. To calculate the roots of the equation, your algorithm should calculate the discriminant D given by

   $$D = b^2 - 4ac$$

   If D > 0, the algorithm displays a message: *The equation has two roots* and then displays the roots.

   If D = 0, the algorithm displays a message: *The equation has one root*, and then displays the root.

   If D < 0, the algorithm displays a message: *The equation has no real roots*.

2. Expanding on Problem 2 from Chapter 6, implement your algorithm. The problem description is restated as follows:

   Write a program that calculates the cost of a car rental according to the following price schedule:

		Rental Period	
Type of Car	**1–6 days**	**7–27 days**	**28–60 days**
Class B	$27 per day	$162 for 7 days, +$25 for each additional day	$662 for 28 days, +$23 for each additional day
Class C	$34 per day	$204 for 7 days, +$31 for each additional day	$810 for 28 days, +$28 for each additional day
Class D	Class D cannot be rented for less than 7 days	$276 for 7 days, +$43 for each additional day	$1136 for 28 days, +$38 for each additional day

   The program asks the user to enter the rental period and type of car. The program then displays the appropriate cost. If a period longer than 60 days is entered, a message *Rental is not available for more than 60 days* is displayed. If a rental period of less than 6 days is entered for Class D, a message *Class D cars cannot be rented for less than 6 days* is displayed.

CHAPTER 8

Iteration and Loops

L OOPS ARE A POWERFUL way to reexecute large portions of code. We can use loops to calculate each time step of a simulation, or to step though the elements of an array or matrix to do element-wise calculations.

8.1 FOR LOOPS

A *for* loop uses a specific loop variable to iteratively run the same block of code a specified number of times. It is typically used to iterate over the elements in an array, either directly or by using a variable to index into the array, depending on language syntax.

8.1.1 MATLAB® Loops

For loop syntax in MATLAB® is very straightforward. In the most basic form, you simply specify a variable, a start point, and an end point. You can use any valid variable name for the loop iterator; we will use j in most of our examples.

```
for j=1:4
        j
end
```

This loop will run four times, displaying 1, 2, 3, and 4 each time respectively, as j is incremented by 1 each time through the loop, starting at 1 and exiting when greater than 4. MATLAB is very quietly creating a vector to iterate over, storing each element in the array in j one at a time as we repeat the loop. This means that we can use MATLAB's vector syntax to precisely control the size that we increment j by each time through the loop. We will commonly want to use the default increment of 1, in order to index in to arrays or matrices, but we could for example step by 0.5.

```
for j=1:0.5:4
      j
end
```

You can nest for loops, which is commonly used to iterate over a matrix.

```
for j=1:4
        for k=1:4
              a(j, k)=j*k;
        end
end
a
```

One important note regarding using for loops to do element-wise operations on arrays or matrices: performance is very poor. If your operation can be done without a loop, using many of the functions and operations in MATLAB that work on arrays, your code will perform much faster.

8.1.2 Python Loops

There are several ways to implement for loops in Python. The simplest example is to use the *range* function to provide a list of integers to iterate over. You can use any valid variable name for the loop iterator; we will use j in most of our examples. Note that indentation is critical in defining the elements to be repeated in the loop.

```
for j in range(4):
        print(j)
```

This loop will run four times, displaying 0, 1, 2, and 3 each time respectively. Although you can use this mechanism to do some operation on every element in an array, Python allows you to iterate directly over a list of items.

```
import numpy as np
a = np.array([1,2,3])
for j in a:
print(j)
```

This example will create the array [1 2 3], and then iterate over this array, printing the first, second, and third items in the array on each run through the loop respectively.

8.2 WHILE LOOPS

Unlike for loops, which have a predefined (even if it is at runtime) number of iterations, while loops iterate for as long as the test condition is true. As soon as the test condition becomes false, the loop exits.

8.2.1 MATLAB® While Loops

The following is a simple example of a while loop in MATLAB.

```
count = 0;
while (count < 4)
        count
        count = count + 1;
end
```

When run, it will output 0, 1, 2, and 3 on each iteration through the loop respectively. When count is equal to 4, the test condition will become false, and the loop will exit.

8.2.2 Python While Loops

The following is a simple example of a while loop in Python.

```
count = 0
while (count < 4):
        print(count)
        count = count + 1
```

When run, it will output 0, 1, 2, and 3 on each iteration through the loop respectively. When count is equal to 4, the test condition will become false, and the loop will exit.

8.3 CONTROL STATEMENTS

Both MATLAB and Python support the use of two special control statements that modify the behavior of a for or while loop.

8.3.1 Continue

The continue command will skip the rest of the commands in the current iteration of a for or while loop, and will cause the code to jump to the next iteration of the loop. In nested loops, it only applies to the loop it is executed in. In other words, a *continue* inside the inner loop will jump to the next iteration of the inner loop, and the outer loop will not be impacted. The syntax for both languages is identical: just the keyword *continue* on its own line.

8.3.2 Break

The break command will cause execution to leave the loop entirely, and the program will jump to immediately after the loop. Just like the continue command, in nested loops only the loop it is executed in is impacted. The syntax in both languages is the keyword *break* on its own line.

EXERCISES

1. Write a for loop that prints all integers from 1 to n.

2. Write a for loop that prints all integers in reverse from n to 1.

3. Write a while loop that prints all even numbers from 1 to 100.

4. Write a while loop that prints all integers in reverse from n to 1.

5. Write a loop that calculates the factorial of n (n!—the product of all positive integers less than or equal to n).

CHAPTER 9

Nonlinear and Dynamic Models

9.1 MODELING COMPLEX SYSTEMS

Although linear models are very useful in analyzing some phenomena and especially static systems, most real systems are much more complex as well as dynamic. For many systems, the relationship between the causes and effects in the model will be nonlinear in form. In addition, the state of the system will change over time and/or space reflecting dynamic changes in the forces underlying the system.

In this chapter, we will examine the components of dynamic systems, extending our approach to creating a conceptual model to reflect that dynamism. We will then review several examples of the mathematical representations of nonlinear systems. Then we will illustrate several mathematical approaches to modeling nonlinear dynamic systems. This chapter then concludes with examples of dynamic systems and the computer algorithms that can used to model those systems.

9.2 SYSTEMS DYNAMICS

Systems dynamics is an approach to the computer modeling of complex systems. It has been used to model any dynamic system "characterized by interdependence, mutual interaction, information feedback, and circular causality" (Systems Dynamics Society, 2016). Jay Forrester from Massachusetts Institute of Technology is often credited with founding this approach to modeling complex systems as part of his first book, *Industrial Dynamics* (Forrester, 1961). In that and subsequent works, he defined an approach to modeling dynamic systems that has become a standard approach to modeling dynamic systems.

9.2.1 Components of a System

As we have already learned, a model is a simplification of a complex system that provides a way to explore the behavior of an explicit portion of that system. For every model we create, we implicitly or explicitly define a system boundary. The relationships we model within the system are said to be endogenous to our model, whereas those outside the boundary are said to be exogenous. Exogenous parameters are often represented as a constant or simulated through a range of values representing the range of circumstances that might occur, but they are not altered by any of the equations endogenous to the model. For example, if we are modeling a regional economy, our boundary is set around an explicit geographic region. The changes in the regional economy will be impacted by the growth or decline of the national economy in which it resides. Those conditions will be exogenous to our regional model and represented by one or more constants.

Within the model, the systems dynamic approach divides the system into several interacting components. The state of the system at any time is measured by one or more levels or stocks at each instance of time. For example, if we are modeling a physical system such as a household heating system, the level we would keep track of over time is the room temperature. For a model of population, we could track the total population and various subdivisions of the population by age, sex, or other socioeconomic characteristics.

Each level interacts with one or more rates. A rate represents the amount of change in a level for each increment of time. Spatial models represent both the rate of change and direction of movement. In natural systems, rates may be governed by physical laws such as the reaction rate of a chemical, the force of gravity acting on an object, or the rate of movement of water through a groundwater aquifer. For business systems models, the rates may represent the decisions made by managers about the business investments or production such as investments in capital equipment, creation and replacement of inventory, or projected sales.

Both the levels and the rates may be limited in their range of values. These can be thought of as constraints imposed either because of resource limitations or physical laws that impose those limits.

The impact of the rates on the levels over time is sometimes referred to as a feedback. A positive feedback results in an increase in the level, whereas a negative feedback reduces the level. A simple furnace and thermostat control is an example of a system with a negative feedback. If the temperature in the room drops below the level set in the thermostat, the furnace is turned on adding heat to the room. Once the desired temperature is reached, the furnace is turned off.

Highways are often cited as creating a positive feedback loop. When there is congestion along an urban corridor, a highway may be built or expanded to relieve that congestion. However, the improved accessibility and short-term gain in traffic movement tends to attract more development. The additional development generates more traffic, which in turn creates additional congestion. One solution to that congestion is the additional expansion of the highway.

Putting these ideas together, we our systems model can be represented mathematically in several ways. One way is to represent the changes in the system as a differential equation. The general form of this equation is shown in Equation 9.1.

$$\frac{dy}{dx} = f(x) \tag{9.1}$$

Simply stated, the rate of change in the variable y(dy) with respect to the change in x(dx) is a function of x. For dynamic models, we are predicting the change in the level with respect to the change in time. The nature of the function f(x) will vary depending upon the phenomenon we are modeling. Later in this chapter, we will look at a variety of examples and create programs to model those changes.

Another way of thinking about rates of change over time is by dividing time into small, discrete intervals. For each time interval, the level we calculate at time t will be the level at the previous time period (t − 1) plus the change in the level that occurred in that interval. The change could be positive or negative, depending on the nature of the relationship. This difference equation can be written as follows:

$$level(t) = level(t-1) + \Delta level \tag{9.2}$$

where:
   level(t) is the level at the current time period
   level(t − 1) is the level in the previous time period
   Δlevel is the change in the level during that time period

For many systems dynamics problems, this is the strategy for modeling systems with a computer. That is why we approximate the change in level that varies depending on the nature of the underlying mathematical function that best represents that particular system. We will review a number of those examples in Sections 9.2.2 and 9.2.3.

9.2.2 Unconstrained Growth and Decay

As the name indicates, unconstrained growth and decay represent a class of models where there are not any constraints on the predicted levels being modeled. Examples of systems that have been modeled in this way are population growth, compound interest on investments, the decay of pollutant concentrations in the environment, and radioactive decay. In these models, the rate of change is proportional to the stock. For each period of time, the amount of growth or decline is a function of the stock times the rate of change.

We can represent the system via a difference equation as shown in Equation 9.3:

$$X(t) = X(t-1) + X(t-1) * R * \Delta t \tag{9.3}$$

where:
  X represents the stock at the current time t and the previous time t − 1
  R represents the rate of change
  Δt is the time increment

A simple example using compound interest is represented in Table 9.1 and Figure 9.1

As shown in Table 9.1, we invest $1000 in year 1. Assuming a 5% annual interest rate, compounded annually, at the end of year 1, the balance is $1050. We then apply the same rate to this new stock in year 2, and then receive $52.50 in interest that year. We carry this forward for 20 years reaching a total of $2526.95.

Figure 9.1 graphs the account balance over time. This type of unconstrained growth is called exponential growth. The exponential function has the form:

$$f(x) = a * b^{kx} \tag{9.4}$$

where:
  a is a scale factor
  b is the base
  k is the growth constant

TABLE 9.1   Account Balance
with 5% Annual Interest

Balance	Year	Interest
$1000.00	1	$50.00
$1050.00	2	$52.50
$1102.50	3	$55.13
$1157.63	4	$57.88
$1215.51	5	$60.78
$1276.28	6	$63.81
$1340.10	7	$67.00
$1407.10	8	$70.36
$1477.46	9	$73.87
$1551.33	10	$77.57
$1628.89	11	$81.44
$1710.34	12	$85.52
$1795.86	13	$89.79
$1885.65	14	$94.28
$1979.93	15	$99.00
$2078.93	16	$103.95
$2182.87	17	$109.14
$2292.02	18	$114.60
$2406.62	19	$120.33
$2526.95	20	$126.35

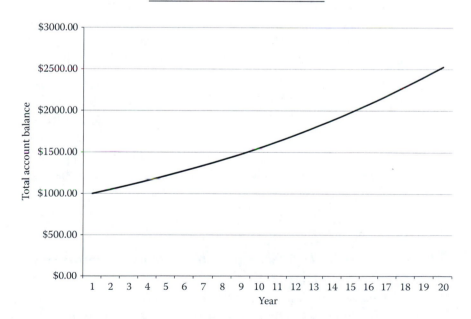

FIGURE 9.1   Account balance for investment of $1000 at 5% annual interest.

If the growth constant is positive, this is a positive exponential, which grows as x increases. If the growth constant is negative, it is a negative exponential, which decreases as x increases. A lot of exponential models use the natural logarithm base e. Some exponential models take the form:

$$x(t) = x_0 e^{kt} \tag{9.5}$$

where:
t is time
the independent variable x is a variable that changes as a function of time
x_0 is the starting value of x
k is the growth constant
e is the natural logarithm base

Exponential functions can be used as a reasonable approach to modeling some systems for a finite period of time. Carried to extremes, exponential growth models can produce ridiculous results. One demographer forecasting the population of the earth showed that after several millennia, the earth would be an expanding ball of flesh expanding outward at the speed of light! Of course, that was admittedly not a forecast he believed. All systems will eventually reach limits to their growth. Populations can only grow to the limits of their food supply or habitat constraints. Decaying substances eventually approach zero or insignificant concentrations. For those reasons, more realistic models add constraints to the growth.

9.2.2.1 Unconstrained Growth Exercises

1. Let us evaluate some exponential functions either in MATLAB® or Python. For the expression y = 4*2^x, generate a set of values for y using values of x from −4 to 10 in increments of 1. Plot a graph of the relationship. Now try varying the constant and observe what happens to the graph. Do the same thing varying the exponential base. Finally, change the exponent to a negative number. How does each change impact the form and range of the graph?

2. Write a program to calculate the interest on an account where you initially save $100 per month over 30 years. Calculate the growth in the account assuming that you receive 1, 3, or 5% interest compounded annually. To keep the calculations fairly simple, only

calculate the interest at the end of each year (having added $1200). Create a graph showing the growth of your savings over time for each of the three interest rates on the same figure. If you want to end up with a minimum of $100,000 in savings after 30 years, how much would you have to change the monthly savings amount to reach that amount at each interest rate? Alter your model to calculate that amount by substituting in different savings amounts in increments of $25 per month. Provide the codes for the original and new solutions along with a one page summary documenting your program and the results.

3. In 1946, Willard Libby and his colleagues were the first to recognize the existence of a radioactive isotope of carbon. The vast majority of carbon has the atomic weight 12, whereas radioactive carbon has the atomic weight 14 (often represented as ^{14}C). This isotope is formed continually in the upper atmosphere by the interaction of neutrons produced by cosmic rays with nitrogen atoms and then becomes part of the carbon dioxide in the atmosphere (Bowman, 1990, p. 10). That carbon dioxide becomes mixed with the atmosphere and is absorbed by all living things. Once a plant or animal dies, it no longer participates in carbon exchange resulting in the radioactive decay of ^{14}C. Libby (1955) is credited with describing how to analyze the ^{14}C to determine the age of archeological and other objects. Although there are a number of assumptions concerning the historical levels of ^{14}C in the atmosphere, its mixing and concentration across the world, and the potential for sample contamination, the simple model of radioactive decay of ^{14}C can be used to approximate the age of any object that was part of the biosphere. Using Libby's estimated half-life of 5568 years, the radiocarbon age of an object is represented by the following equation:

$$t = -8033 \ln \frac{A}{A_0} \tag{9.6}$$

where:
  t is time
  A/A_0 is the ratio of the remaining ^{14}C atoms to the amount in the atmosphere
  ln is the natural logarithm

Write a program to create a graph of the number of atoms remaining in a sample versus time. What is the approximate age of a bone found with 0.77% of ^{14}C?

9.2.3 Constrained Growth

The obvious problem with models of unconstrained growth is that physical, ecological, and human systems are limited or constrained by the availability of energy, resources, or other capacity limitations of the system. For businesses, the amount of growth will be limited by things such as the availability of capital, the current production or staff capacities to build and deliver products, competition from other providers, and market saturation. Similar constraints will apply to all sorts of human-managed systems like public transit, public housing, and health systems.

For natural systems, forecasts of population change may be limited in many ways. Of course, all populations experience deaths from natural and other causes. As the population increases, the food supply may become limited, raising the death rate in the population and therefore altering the population base total. Food shortages may also alter the fertility rate of the population, essentially changing the original growth rate. Other possible constraints include disease, habitat alterations, predation, limited energy resources, and limits to other nutrients. Extreme stresses on the population such as major habitat losses, over harvesting by humans, and extreme weather events may cause a complete population collapse. Meadows (1972), using the methods introduced by Forrester, published a famous global look at these environmental limitations in the book *The Limits to Growth*. Although technological advances in agriculture and in other sectors have rendered some of their predictions invalid at the global scale, there are certainly regional scale problems that have emerged due to shortages of water, fertilizer, and arable land.

There are several ways to reflect such constraints in systems models. One approach is to incorporate a mathematical function that mimics the shape of the growth curve over time as the population approaches these limitations. In population forecasts, the logistic curve is often used in this way. The general form of this equation is shown as Equation 9.7.

$$f(x) = \frac{L}{1 + e^{-k(x-x_0)}} \tag{9.7}$$

where:
    L is the curves maximum value or limit
    k is the steepness of the curve

x_0 is the value of x-value of the midpoint
e is the natural logarithm

Assuming there are no significant changes in the environment, the maximum value of the population can be defined as the carrying capacity for that population. The carrying capacity is the maximum size of the population that can be supported by a particular environment. One way of implementing this function in a population forecast is to add a death rate function that subtracts from the population as it approaches the carrying capacity.

Putting this together with the exponential growth rate associated with births results in the following equation:

$$\Delta P = P(t - \Delta t) * R * 1 - \frac{(P(t - \Delta t))}{C} * \Delta t \qquad (9.8)$$

where:
ΔP is the change in the population
$P(t - \Delta t)$ is the net population change in the previous period
R is the birth rate
C is the carrying capacity

The first term represents the results of the births in the period added to the previous period base. The second term uses the ratio between the resulting population and the carrying capacity to modify the additional population added in the time increment. When the ratio is small, the population will grow at nearly the exponential rate. As the carrying capacity is approached, there are increasing reductions in the population resulting in a leveling off of the population growth. One of the exercises for this chapter is to program that model and graph the resulting distribution.

Although this approach may mimic the form of the population changes, the model does not explain the underlying causes of those changes. More sophisticated models have explicit relationships across the growth and constraint factors. In population models, these may include age-specific death rates, the limitations of the food supply or habitat, the occurrence of disease, predation, and other factors.

In the industrial dynamics models started by Forrester, constraints included such factors as the availability of capital, the current production capacity, market penetration, availability of labor, and management expertise.

Similar approaches have been used to forecast a variety of social and economic phenomena including land use change, the growth of social movements, and the housing market.

Estimating the values of all of the relevant parameters can be difficult. The accuracy of the model will be influenced by the availability of data for the system in question. However, even without exact measurements of all of the relationships, models can be used to answer *what if* questions associated with reasonable variations in the model parameters, providing important insights into system behavior.

9.2.3.1 Constrained Growth Exercise

To model constrained growth, you will build a population dynamics models for predator and prey, which includes part of the interactions conceptualized under this topic. The interactions are based on a model formulated by an American demographer Alfred Lotka and an Italian physicist named Vito Volterra working independently (see a summary of this work at https://en.wikipedia.org/wiki/Lotka%E2%80%93Volterra_equations). The model expressed as a difference equation for rabbits and wolves is as follows:

$$\text{Rabbits} (t+1) = \text{Rabbits} (t) + \text{rabbits} (t) * \text{rabbit growth rate}$$

$$- \text{rabbits} (t) * \text{rabbit death rate} * \text{wolves} (t)$$

$$\text{Wolves} (t+1) = \text{Wolves} (t) + \text{wolves} (t) * \text{wolf growth rate} * \text{rabbits} (t)$$

$$- \text{wolves} (t) * \text{wolf death rate}$$

where t is the time increment.

If we start with the basic growth model from interest and instead calculate rabbits, the only information we need is the rabbit growth and death rates and the initial number of wolves to satisfy the first equation. Similarly, we can use the basic growth rate model to model the wolf population needing only the wolf birth and death rates and the initial rabbit population. Create a model that incorporates these new factors and plot the population of rabbits and wolves on the same graph over time.

Here are the parameters you should use to start. If your model works correctly, you should observe a cyclical pattern where the prey

TABLE 9.2   Initial Parameters for Predator Prey Model

Parameter	Rabbits	Wolves
Growth rate	0.1	0.005
Initial population	40	15
Death rate	0.01	0.1
Time period	200	200

increases followed by an increase in predators and then a decline in both (Table 9.2).

You should notice that the pattern that emerges is becoming increasingly unstable, with larger peaks and valleys. Change the model parameters to investigate how a more stable population might be achieved. Then discuss the design of a more complex model including the incorporation of additional components of the data that would be needed to implement it. Discuss the findings of your model along with those concerning the design of the more complex model.

9.3 MODELING PHYSICAL AND SOCIAL PHENOMENA

The growth model examples illustrate how one implements a nonlinear model given a set of governing equations that represent changes in the phenomenon being modeled with respect to time. Models of physical phenomena use scientific laws and theories to construct the equations governing their behavior. Those theories have also been tested against experimental data that have confirmed the efficacy and have led to modifications of the representation that more closely match the experimental evidence.

For social phenomena, there are also underlying theories but human behavior is not subject to the definition of scientific laws with the same level of certainty. Instead, those models are most often based on empirical evidence from primary or secondary data sources that are used to estimate one or more governing equations. Making such estimates will be discussed further in Chapter 10.

In either case, anyone modeling a system needs to understand its limits may lie either in empirical terms from observations of data or from the underlying scientific or engineering principles that have been developed from accumulated knowledge. Following these general guidelines will help to ensure that the models you formulate either account for those limits or explicitly indicate that they are missing and need to be considered:

1. Define the purpose of your model and the accuracy required to make a *correct* decision.

2. Conduct a literature search that focuses on models of the same phenomena that have been developed by others. Full understanding of more advanced models will often require a deeper background in mathematics as models are likely to use linear algebra, differential equations, and/or partial differential equations to express the underlying relationships.

3. Create a conceptual model that accounts for the factors that are included in those other models and the assumptions made in those models.

4. Identify how well the models estimated the observed results and the potential sources of error indicated in the literature.

5. Identify the risks associated with your model giving a wrong answer and how this relates to the known errors. For example, if you are modeling a skydiver, an overestimate of how effective the parachute will be at slowing down the decent once it is open could result in injury or death. In such a case, we would prefer to err on the conservative side and either open the parachute sooner or use a bigger parachute that slows us down more to avoid that risk.

6. Honestly report the limitations of your model with respect to both known and unknown causes of model variation.

If you take these steps, you will build an understanding of the system being modeled and will build a model that provides the insights you seek.

9.3.1 Simple Model of Tossed Ball

Suppose we are standing on a bridge and toss a ball into the air over the side of the bridge. To model the vertical fall of the ball until it reaches the ground, we must account for several forces acting on the ball. When we toss the ball, we are exerting an upward force on the ball. At the same time, the force of gravity is exerted in the opposite direction, which, after a short time, results in the ball falling toward the ground. To track the progress of the ball, we can create a model that includes its velocity, acceleration, and position over time. At any instant of time, the velocity can be derived by calculating the change in position over the time period. Thus velocity can be represented as

$$v(t) = \frac{dp}{dt} \tag{9.9}$$

where:
   v is the velocity at time t
   dp is the change in position

Acceleration (a) is then the change in velocity over time:

$$a(t) = \frac{dv}{dt} \tag{9.10}$$

The change in acceleration is due to the force of gravity. This is approximately -9.81 m/s².

These relationships lead to the derivation of these kinetic equations for the free fall of an object due to gravity:

$$p = v_i * t + 0.5 \, gt^2 + p_0 \tag{9.11}$$

$$v_i = v_0 * g^t \tag{9.12}$$

where:
   p is the position of the ball and p_0 is its initial position
   v_i is the velocity and v_0 is the initial velocity
   g is the acceleration due to gravity
   t is the time

These equations are used to solve for the position and velocity of the ball in one of the programming exercises in this section.

9.3.2 Extending the Model

The model of a free falling object assumes that there are no other forces acting on the object. In order to account for friction and other forces that act on such an object, we need to incorporate several other components. The first of these is Newton's second law of motion.

Newton's second law states that a force F acting on an object of mass m gives the object acceleration a or

$$F = ma$$

Force is measured in N or kg m/s². The mass is measured in kg and the acceleration as m/s². The acceleration is due to the force of gravity as used in the ball toss model.

The other force acting on a falling object is friction. For an object moving through a fluid, the frictional force is called drag. In the case of the ball, a bungee jumper, or skydiver, the objects are moving through air. The general formula for the force of drag F_d is as follows:

$$F_d = \frac{C_d \rho v^2 A}{2} \tag{9.13}$$

where:

C_d is the drag coefficient
ρ is the density of the fluid; for air at sea level this is 1.29 kg/m³
v is the velocity of the object
A is the objects area in the direction of movement called the characteristic frontal area

The drag coefficient is a dimensionless coefficient that takes into account the object's shape. Note that the drag increases with the square of velocity. The drag force is in the opposite direction of the force of gravity. Thus, for a skydiver, the parachute is designed to create sufficient drag so that the person will hit the ground at a speed that will not cause any injury.

A more complete model of a skydiver will therefore need to account for the forces in both directions and how they vary with respect to time in order to track the velocity of the skydiver as they reach the ground. Creating such a model is one of the capstone projects presented in Chapter 14.

9.3.2.1 Ball Toss Exercise

Using the information in this chapter, program a simple model of a ball tossed straight up from a bridge. Model it as a falling object as discussed in Section 9.3.1. Assume that the bridge is 15 m above the ground and that it is thrown upward at a velocity of 25 m/s. Model its velocity and position over time until it reaches the ground. Discuss the following questions as part of your exercise:

1. What are the assumptions underlying this model?

2. What other functions would we need to add to the model to relax the major assumptions?

REFERENCES

Bowman, S. 1990. *Radiocarbon Dating.* Berkeley, CA: University of California Press.

Forrester, J. 1961. *Industrial Dynamics.* Cambridge, MA: MIT Press. (Also available online at https://babel.hathitrust.org/cgi/pt?id=mdp.39015002111774;view=1 up;seq=9).

Libby, W. F. 1955. *Radiocarbon Dating.* 2nd ed. Chicago, IL: University of Chicago Press.

Meadows, D. H. 1972. *The Limits to Growth: A Report for The Club of Rome's Project on the Predicament of Mankind.* New York: Universe Books.

Systems Dynamics Society. 2016. Introduction to systems dynamics. http://www.systemdynamics.org/what-is-s/ (accessed August 2, 2016).

CHAPTER 10

Estimating Models from Empirical Data

10.1 USING DATA TO BUILD FORECASTING MODELS

There are a number of circumstances when there will not be an existing modeling framework and mathematical representations of the system under study. For example, one may be working with a new material whose responses to different stresses are unknown. In such a case, there may be experimental data that describe the relationships between the state of the material and the stresses applied. Scientists who are exploring the relationships between genetics and disease mine those data to define and test models of those relationships. In a similar way, data on the economy are used to derive models of market behavior under a variety of circumstances.

In all of these cases, one must use the available data to build an empirical relationship among the causes and effects that can then be used to simulate the behavior of the system. There are a large number of statistical techniques that can be used to build such relationships. A comprehensive presentation of those techniques is beyond the scope of this chapter. For those interested in further exploration, there are some references available at the end of this chapter. Instead, we will focus on introducing the techniques for fitting a function to an empirical dataset, estimating the goodness of the fit, and discussing the limitations and pitfalls of incorporating those functions into a predictive model.

10.1.1 Limitations of Empirical Models

Models based on empirical relationships can be a valuable starting point to understanding how systems behave. However, there are a number of assumptions that are implicit in creating models from those relationships that need to be taken into account.

For data obtained through laboratory experiments, there are generally clear cause and effect relationships embedded in the experimental design. Under a set of standard conditions, the experiment varies the inputs (causes) in a consistent way and then measures the outcomes (effects) on the experimental system. A plot of the causes versus effects can be used to define a mathematical function that best fits the system. There will be several limitations to the application of that function to forecasting the outcome under circumstances that were not measured in the laboratory. First and most important, the forecast will not be valid beyond the limits of the experimental data. The behavior of the system beyond the experimental results is unknown. Although it may continue along the same trajectory, it might also completely change its behavior under different circumstances.

> An empirical model cannot be used to predict the system behavior beyond the range of the data.

Caution must also be used because the experiments have been run under a standard set of conditions and have not tested the relationships under all possible conditions. For example, tests of the mechanical properties of a new material may not have included all possible types of mechanical failure or not tested for failure under all possible circumstances. Behaviors may change under different pressure or temperature conditions. The empirical data describes a limited set of relationships but does not fully explain the relationships in the entire system. Thus, the limitations of the experiment translate into limitations in the models created from the experimental data.

For datasets that include observations in open systems where standardized controls are not possible, additional limitations arise. For example, samples of aquatic biology and related physical and chemical measures of water pollution are only taken intermittently in very discrete locations along a stream. Sampling then has both spatial and temporal limitations that could limit the ability to explain a decline in the diversity of the aquatic biota in response to environmental stresses such as water pollutants. The response of the biota to a stress may be due to an event in the same section of the stream or an upstream event, which migrated downstream. Depending on the timing of water sampling, the cause may or may not be captured in the dataset. In addition, the cause could be a combination of several pollutants, some of which might have been measured and others that have not been measured. An empirical relationship in such a circumstance may capture one or more independent variables that appear to be related to the decline in biota but there may be other causes that covary with what has been measured. In addition, the sample may not be representative of all possible conditions, which will introduce other biases into the statistical analysis.

These kinds of problems can be even more complicated with socioeconomic data gathered by the U.S. Census or through other surveys and indices. Individual responses are generally not available to the public to protect the privacy of individuals and businesses. Thus, they are reported as summaries for larger geographic units that are not necessarily homogenous in their measured characteristics. A census tract may encompass several different neighborhoods with a different mix of population, housing, public services, and businesses. In addition, many socioeconomic variables are highly correlated with one another making it difficult to define what is the real cause of a change. For example, a data analysis of migration within a metropolitan area from the central city to the suburbs might show that both income and education are the important determinants of the likelihood to migrate. However, income and education are highly correlated. Do people migrate because they have the economic means to move and desire to have larger single family homes in the suburbs? Are they more educated and therefore desire to move because they perceive that suburban schools are of higher quality? Is there another circumstance that is the underlying cause of their migration such as the perception of crime in the target neighborhoods that is not represented in the data? All of these may play a role but sorting them out from the empirical data may not be possible. One must carefully think through the possible causes and effects and use one or more proxy variables that are representative of the possible driving forces to forecast the trends without placing too much confidence that the causes are understood.

> Empirical models can be used to forecast the trends in circumstances but do not necessarily fully capture the underlying causes of the changes in the system.

Finally, the sampling associated with empirical models constrains the validity of the projections if there are circumstances that would change the relationships if the samples were taken at a different time or in different places. Going back to our example of the diversity of biota in a stream, we would need to ascertain whether the datasets are representative of a *typical* year or if they contain instances of infrequent but cyclical events such as droughts or floods. The stresses caused by these intermittent events can easily create a bias in our empirical model. In a similar vein, the spatial distribution of our sampling distribution may not capture unusual events in the watershed such as a toxic spill or the runoff of a large amount of sediment from a construction site upstream.

With socioeconomic data, we often use both current surveys and historical data to construct models of a myriad of behaviors relating to markets, travel, and social trends. These models assume that our past *samples* are representative of what will happen in the future. However, there may actually be changes in technology, economic circumstances, or social attitudes that profoundly change what will happen in the future. If someone was modeling the underlying causes of traffic fatalities, could they have anticipated the change in technology that has caused a large number of accidents because of the distractions of text messaging?

> When using an empirical model to make forecasts for a system, it is important to specify all of the implicit and explicit assumptions associated with the spatial and temporal nature of the underlying dataset.

With all of these caveats in mind, it is still useful to build models from empirical datasets. Such models can lead to a deeper understanding of the system and guide future research, policy, and management decisions. The relationships that are defined may also lead to a full explanation of system behavior that can be incorporated into future models. The following sections introduce methods to define and measure the efficacy of empirical models and provide several examples.

10.2 FITTING A MATHEMATICAL FUNCTION TO DATA

The first step in creating a model from empirical data is to assess the nature and strength of the relationship between the independent (causal) variable(s) and dependent (effects) variables. This can be done in several ways. For a simple two variable case, plotting the data points for one variable against the other (often called an XY or scatter graph) may be good starting point.

As an example, download the file paintcr2.csv from the book website. The file has two columns representing a sample of two water quality variables from the Paint Creek watershed in Ohio. The first column is a numeric indicator of the diversity of the fish population in a section of the creek called the index of biotic integrity (IBI). The IBI ranges in value from 12 to 60 where values in the range from about 44 to 60 indicate watersheds that exhibit the diversity of the fish population that would be expected in an unpolluted, natural stream in Ohio. The second column represents the maximum value of total suspended solids (TSS_MAX) found in the same stream segments where the IBI was measured. Sediments are pollutants brought into the stream via erosion that damage the aquatic habitat by covering the bottom of the stream where fish may lay their eggs or find their food sources. At the extreme, sediments make it difficult for fish to see their prey, further hindering their success. Thus, we would hypothesize that as the level of TSS goes up, the diversity of the fish population would go down. We will use the data to test this hypothesis and create a statistical model of that relationship. Use the following instructions for MATLAB® or Python to import the dataset and make an XY plot of the relationships:

> For those using MATLAB, from the home tab, choose Import Data. A window will open that will allow you to go to the directory where you have downloaded the file and choose the paintcr2.csv file. The table should appear showing the nine rows representing the samples for IBI and TSS_MAX. Relabel the columns by double-clicking and inserting the correct variable names. Then click on the import selected button to import the data. Use the code in the following box to create an XY graph.

> For Python, click variable explorer and choose the Import Data option from the right side icons. Navigate to the paintcr2.csv file and import it as data. It will appear in the variable explorer as a 9 by 2 matrix. Use the Python code below to extract the two variables and make the XY plot.

MATLAB CODE

```
sz=45;%sets the size of circles in the plot
scatter (TSS_MAX, IBI, sz,'ko','filled');%x axis is TSS
black circles
```

PYTHON CODE

```
import math
import matplotlib.pyplot as plt
import numpy as np
IBI=np.zeros(9)
TSS_MAX=np.zeros(9)
IBI=paintcr2csv[0:9,0]
TSS_MAX=paintcr2csv[0:9,1]
x=TSS_MAX
y=IBI
plt.scatter(TSS_MAX, IBI)
```

You should see a plot that matches Figure 10.1.

10.2.1 Fitting a Linear Model

Examination of the scatterplot shown in Figure 10.1 indicates that the relationship between these two variables appears to be linear. We could draw a line on the graph that comes near many of the points but that would be arbitrary and would not yield a measure of how well that line compares to other possible lines.

The standard way of tackling this problem is using the technique called linear regression. Linear regression assumes that there is a cause and effect

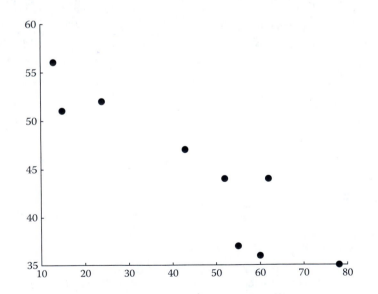

FIGURE 10.1   X-Y plot of index of biotic integrity and maximum total suspended solids.

relationship between the predictor or independent (x) and response or dependent (y) variable:

$$y = a + bx + \varepsilon \qquad (10.1)$$

where:

x and y are the predictor and response variables respectively

a is the intercept of the line

b is the slope of the line

ε is the error or residual between the model and the observations

The approach used to establish the best fit line and calculate the error is to minimize the sum of the squares of the residuals. This is illustrated graphically in Figure 10.2. The residuals are measured as the vertical distance between the line and the observations and several of which are shown in the figure as dotted lines. For our example dataset, we provide an exercise at the end of this chapter to derive the relevant regression equation.

We are assuming at least a basic understanding of the general principles of descriptive and inferential statistics and will not repeat them here. In addition, the underlying mathematics of the least squares function are illustrated in great detail by several authors and will not be covered here (Draper and Smith, 1998; Chapra, 2008). Please consult these references and any introductory statistics sources for further information and more extensive applications of regression techniques.

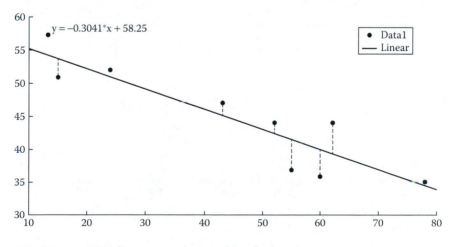

FIGURE 10.2   Fitted regression line and residuals.

For our use in building a linear model, we need to understand several important statistics that are calculated as part of regression analysis. First, there is a standard error of the estimate. This is the measure of the dispersion of the original data around the regression line. The second statistic is called the coefficient of determination or R squared (R^2). This number represents the percentage of the variance that is explained by the regression equation. The closer the R^2 value is to 1, the better the fit between the sample data and the regression line.

The third set of statistics involves a test of the significance of the regression equation and its coefficients. We want to make sure that the relationship could not have happened by chance. Two types of tests are often applied in this regard. For the equation as a whole, a comparison is made with a standard table called the F distribution. Both the F statistic and the probability of F are often reported by statistics program. We want to be able to say that there is a low probability that the relationship happened by chance. For the most stringent test of significance, we want to find that the equation had less than a 1% probability of happening by chance. Sometimes a probability of less than 5% is also acceptable. The t test is a similar test of probabilities associated with both the intercept and the coefficient for the slope of the line. These are again reported as a probability that is compared with the 1% or 5% criterion to determine the significance of the coefficients.

Table 10.1 is an illustration of the full output of a linear regression routine. The dataset used is from the Center for Disease Control of statewide rates of smoking and heart disease deaths. The percentage smokers is the predictor variable and the heart disease deaths are the response variable.

TABLE 10.1 Regression Results from CDC Data on Smoking and Heart Disease Deaths by State

Model	Linear Regression Model: Rate ~ 1 + CurrentSmoking			
Estimated coefficients:	Estimate	Standard Error	tStat	pValue
(Intercept)	40.672	22.776	1.7857	0.080332
CurrentSmoking	5.1244	0.97289	5.2672	3.0752e-06

Number of observations: 51, Error degrees of freedom: 49

Root mean squared error: 20.4

R-squared: 0.362   Adjusted R-squared 0.348

F-statistic vs. constant model: 27.7   pValue = 3.08e-06

The table was generated using the MATLAB fitlm (for fit linear model) routine that is in the statistics toolbox. There are actually several ways in both MATLAB and Python to generate regression and curve fitting functions, which will be discussed later in this chapter.

We can use the output shown in Table 10.1 to construct our linear model and judge its efficacy. The coefficients for the linear model are shown under the estimate column in the table. If we accept the model the equation would be as follows:

$$\text{Heart Rate Deaths} = 40.672 + 5.1244*\text{CurrentSmoking}$$

We can see that the pValue for the intercept is over 0.05, and the value for CurrentSmoking is much less than 0.01. Thus, we might want to run the model again forcing the intercept to be zero. We also see that the overall model is significant with a very small pValue but we are only explaining part of the variance in the distribution. There are two R-squared values that show us this. The adjusted R-square is usually the one we use. It is adjusted relative to the degrees of freedom associated with the model. It shows us that 34.8% of the variance is explained.

Clearly, there are other risk factors for heart disease deaths that are not captured by this empirical model. We could choose to use this model, acknowledging that it underestimates the risk. Alternatively, we could search for other risk factors that might create a better model.

10.2.2 Linear Models with Multiple Predictors

When we have data with several possible predictors associated with a response, we can use multiple linear regression to estimate the contribution and significance of each of the predictor variables. We do this by combining several linear estimators in the same equation as represented in Equation 10.2:

$$y = a + b_1x_1 + b_2x_2 + b_3x_3 + \dots b_nx_n + \varepsilon \qquad (10.2)$$

The solution provides a coefficient b for each of the x predictor variables along with an indication of their significance. There are several circumstances, which may produce a biased estimate of the coefficient of determination and coefficients. One occurs when two or more of the predictor variables are strongly correlated or colinear. For example, this often occurs in social science models that use demographic information such as age, income, and education, which are strongly related. There are several other statistics that can be used to test for this problem and follow-on strategies to remove the biases. A full discussion of these techniques is beyond the scope of this chapter. Draper and Smith (1998) discuss all of these problems and strategies in detail as do many statistics textbooks.

In multiple regression, the output will provide a table showing the significance of each of the independent variables, the R^2 values, and the significance of the overall equation. In this way, we can determine which of the predictor values to use in a final empirical model while getting an estimate of the overall variance explained by the model and the scope of the errors.

10.2.3 Nonlinear Model Estimation

What do we do if the relationships between the predictor and response variables are nonlinear? In such cases, we have several choices for creating an empirical model. In some cases, we can apply a mathematical transformation of the input data that create a linear representation of the system. Table 10.2 shows the transformations for the exponential and power functions, their linear form, and the predictive equation that can be used to make a forecast with the estimated coefficients. For the exponential distribution, we take the natural log (shown as ln) of the response variable.

The coefficients from that analysis can then be used with the original data as a predictor of the response variable y as shown under the predictor equation column in Table 10.2. Similarly, the power function can be linearized by taking the logarithm (here shown as log for log base 10) of both the predictor and response variables and then apply the linear coefficients to create a prediction equation.

As an example, we will use the data from the graphing example in Chapter 5, which showed the reduction in light intensity with distance. Figure 10.3 shows that relationship with a sample of seven measurements.

We can recognize from the shape of the graph that this is probably a negative exponential function. In order to linearize the relationship, we will need to take the natural log of the response variable, light intensity

TABLE 10.2   Examples of Linear Transformations of Nonlinear Data

Distribution	Equation	Linear Form	Predictive Equation
Exponential	$y = \alpha e^{\beta x}$	$\ln(y) = \ln(\alpha) + \beta x$	$y = e^{\alpha} + \beta x$
Power	$y = \alpha x^{\beta}$	$\log(y) = \log(\alpha) + \beta \log(x)$	$y = 10^{\alpha} x^{\beta}$

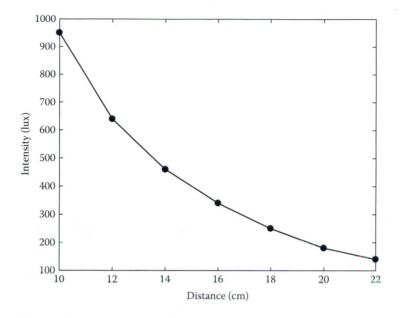

FIGURE 10.3   Light intensity as a function of distance from the source.

and then regress this against the predictor variable, distance. Using either MATLAB or Python, import the raw data from the file light_data.csv from the book website. The first column is the predictor variable, and the second column is the response variable.

Our first step is to create a new variable *llight*, which is the natural logarithm of the predictor variable. To find the built-in functions for this, you can use the help item in MATLAB to search for math functions. For the Math and SciPy modules, you can find a list of the built-in math functions on the websites cited at the end of this chapter. In both cases, this is the log() function. Thus, we take the log of the response variable vector to create a new vector called *llight*.

Now we can use one of the tools from MATLAB or Python to create a linear regression of *llight* as a function of the original distance data. In both cases, there are several options that can be used to fit a linear model. They vary both in their syntax and the amount of information they give us about the goodness of fit.

Tables 10.3 and 10.4 provide examples of the several ways that one can fit a linear or nonlinear equation to a dataset in MATLAB and Python, respectively. Here, we will provide one example of each but provide an opportunity to try several others as part of the exercises at the end of this chapter.

TABLE 10.3   MATLAB® Procedures to Fit a Linear or Nonlinear Model

Procedure and Syntax	Description
Curve Fit App in the GUI interface	Allows a fit of two variables for both linear and nonlinear functions. Linear regression is fit using a first-order polynomial and outputs a graph, coefficients, and R-squared values.
curve_fit function: curve_fit=fit(x,y,fittype)	Using the fit type, "poly1" produces the same statistics as the app but without the graph. Choosing exponential or another nonlinear form will fit that form to your data.
fitlm: lim=fitlm(x,y)	Use in program or command line. Provides coefficients, t and F statistic probabilities, R-squared values, residuals, and other statistics.

fitnlm:
beta0=[5000−0.1];
X=Distance;
y=Lumens;
modelfun= @(b,x)(b(1)*(exp(b(2)*x)));
mdl=fitnlm(X,y,modelfun, beta0)
	Fit a nonlinear model to a nonlinear function provided by the user along with an initial estimate of the coefficients. Returns the coefficients along with the t and F statistics probabilities. Requires an initial estimate of the coefficients that is reasonable, or the algorithm may not work.

TABLE 10.4   Python Procedures to Fit Linear and Nonlinear Models

Procedure and Syntax	Description
Scipy.optimize.curve_fit from scipy.optimize import curve_fit def line(x, a, b): return a * x + b popt, pcov=curve_fit(line, x, y)	Defines a linear or nonlinear function and then input function, x and y variables to curve_fit. Returns the coefficients and covariance matrix.
Scipy.stats.linregress from scipy import stats slope, intercept, r_value, p_value, std_err = …stats. linregress(x, y) print (intercept, slope, r_value**2, p_value)	Linear regression that returns coefficients, r, and p values.
Statsmodels ordinary least squares[a] import pandas from statsmodels.formula.api import ols data = pandas.DataFrame({'x': distance, 'y': llight}) model = ols("y ~ x", data).fit() print (model.summary())	Full function ordinary least squares. Requires the use of the panda dataframes. Returns all the coefficients, p values, and a number of other statistical measures relating to the equation.

[a]   There are a number of related routines in NumPy, Scipy.stats, and statsmodels related to these examples.

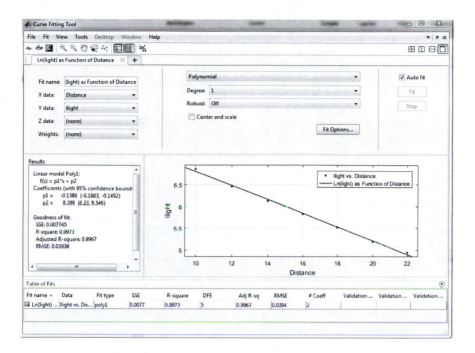

FIGURE 10.4   Output of the MATLAB® curve fitting app for the light attenuation data.

For MATLAB, we will use the built-in curve fitting application to fit the transformed data to a linear equation. Go to the top menu in MATLAB with a tab labeled as APPS and choose Curve Fit. Add a title and choose the input values for x (distance) and y (llight). The standard linear model can also be represented as a first degree polynomial. Thus, in the center of the curve fit window, we choose polynomial, degree 1 as the fit and click on the Fit button. You should get a screen that looks like Figure 10.4. The graph shows the predicted line with a plot of the points around that line. The left side text shows the regression results.

We can perform a similar function in Python using the Scipy curve fit option shown in Table 10.4. For many of the Python and several of the MATLAB options, we need to specify the function that will be used to fit the data. As shown in Table 10.4, we define a linear function and then use this function in the curve fit procedure. The function returns the coefficients in the linear equations. We could then use the plotting functions to create a graph similar to Figure 10.4.

10.2.3.1 Limitations with Linear Transformation

Transformation does not always provide the best model for a nonlinear distribution. The regression model assumes that the scatter of points around the best fit line is a normal distribution. This is not always the case with the transformed data. What were outliers in the original distribution may appear to be closer to the fitted line and bias the R^2 value. Care should be taken to calculate the predicted values and plot them with a scatterplot of the original data to see if there is a good match.

10.2.3.2 Nonlinear Fitting and Regression

A second choice is to fit a curvilinear function to the original data. Fitting to a curvilinear function follows the same general principles for calculating the goodness of fit. In nonlinear regression and curve fitting, a nonlinear equation is fitted as part of what is called the generalized linear model. Both MATLAB and Python have algorithms that then find the least squares fit for the dataset. Tables 10.3 and 10.4 show some of the nonlinear fit methods for MATLAB and Python.

The curve_fit functions in MATLAB provide a choice of fit types that can be used in the calculations. The fitnlm (for fit nonlinear model) requires the user to input a function that is called in the function syntax. The model shown in Table 10.3 uses this syntax: modelfun= @(b, x) (b(1)*(exp(b(2)*x))). The variables in the model are b and x, the intercept and independent variable, respectively. The mathematical representation that follows defines the exponential model with the two coefficients: b(1) and b(2). Other mathematical functions such as third degree polynomials would have additional coefficients for which to solve. There are also functions that allow more than one independent variable in the generalized linear model. Those models are not shown here but follow the same principles except that more than one independent variable is specified by providing a matrix of inputs.

A similar approach is taken in the Python examples as shown in Table 10.4. Functions are used to define the form of the nonlinear fit and then submitted to the curve fit function. Python also has several statistics packages. These include the Scipy.stats and the Statsmodel packages (Scipy.stats, 2016; Statsmodels, 2016). The most robust generalized linear model is in the Statsmodel package. It is often coupled with the use of the Pandas Data Analysis Library, which facilitates the handling of large datasets and provides a simplified syntax for specifying statistical models (Pandas Library, 2016). In Table 10.4 we show the simple syntax for the ordinary least squares made possible by importing the panda library. We will not be reviewing that library here but provide a reference to those who wish to use it (Thomas, 2016).

10.2.3.3 Segmentation

In circumstances where the data distribution has two or more distinct shapes, an approach called segmentation can be used to create separate linear or nonlinear models for each segment. Such circumstances might arise when the phenomenon being modeled changes behavior at distinct points. For physical phenomena, that may be the case if there is a change in state due to changes in conditions like temperature or the amount of stress. For social systems, there may be a tipping point where a market changes behavior such as the real estate values in a gentrifying neighborhood or rents in parts of cities with high crime rates.

The same curve fitting approaches are followed for each of the distinct segments. The resulting model then must check for the values of the independent variables at the segment boundaries and apply the appropriate predictive model for that segment.

EXERCISES

1. Use the Paint Creek dataset to find the coefficients for a linear model using at least two of the different procedures in either MATLAB or Python. For the procedures that do not produce a graph, create a graph showing the regression line and the distribution of the original points around that line. Create a report showing your results.

2. Use the light dataset to create both the transformed linear model and a nonlinear, exponential model of the relationships in either MATLAB or Python. For the linear model, calculate the predicted values from the model and plot that as a line along with the scatter of points from the original dataset. For the nonlinear model, create a similar graph. Compare the two models in relation to the values of adjusted R^2, the significance of the coefficients, and the scatter of the original data around the predicted curve. Which model do you think works better in this instance? Create a report showing your results and analysis.

3. Find another dataset of interest to you where you hypothesize a cause and effect relationship between one response variable and one or more predictor variables. Plot the relationships and then choose

one or more curve fitting functions to test the strength of the relationships. Prepare a report showing the statistical results and plots illustrating the model.

FURTHER READINGS

Math Functions. Scipy.org. https://docs.scipy.org/doc/numpy/reference/routines.math.html#exponents-and-logarithms (accessed November 11, 2016).

Mathematical Functions. Python Software Foundation. https://docs.python.org/3/library/math.html (accessed November 11, 2016).

REFERENCES

Chapra, S. C. 2008. *Applied Numerical Methods with MATLAB for Engineers and Scientists*. New York: McGraw-Hill.

Draper, N. R. and H. Smith. 1998. *Applied Regression Analysis*. New York: Wiley.

Pandas Data Analysis Library. http://pandas.pydata.org/ (accessed November 11, 2016).

Scipy.stats Documentation. https://docs.scipy.org/doc/scipy/reference/stats.html (accessed November 11, 2016).

Statsmodels Documentation. http://statsmodels.sourceforge.net/stable/index.html (accessed November 11, 2016).

Thomas, H. 2016. *An Introduction to Statistics with Python*. Switzerland: Springer International Publisher.

CHAPTER 11

Stochastic Models

11.1 INTRODUCTION

Thus far, we have been working with deterministic models. Those models have used parameters that remain constant throughout the simulation, leading to a singular result. We have used sensitivity testing to explore the impacts of changes in selected parameters but always in the same deterministic framework.

There are many situations in which the values of one or more of our model components are uncertain. There is an element of chance associated with their values. In such cases, the models are said to be probabilistic or stochastic.

> A stochastic or probabilistic model is one, which includes one or more random variables. Such models use Monte Carlo or random experiments to draw values for selected variables from a probability distribution or sampling of an empirical distribution.

There are many examples of situations where a Monte Carlo simulation would better represent the system being studied:

- Environmental models impacted by variations in weather conditions

- The Brownian movement of molecules in a solution

- The behavior of a market

- Risks associated with business investments

- The spread of a forest fire with respect to wind speed and direction

- Fluid dynamics simulations

For some of the simple systems we have used as examples earlier in the book, we can imagine stochastic versions of the models to represent those systems. For example, our traffic model, we could insert a random probability that we encounter other cars or pedestrians at intersections that would add time to our commute. We could add the impacts of weather events such as precipitation that would slow traffic below normal speeds. In fact, in our community of Columbus, Ohio, the mere prediction of snow slows down the traffic!

In each case, we would create a random variable based on a sample of data from observations to represent the probabilities of different conditions. A random number generator would then be inserted into our code to simulate the random effects. We would then run the model many times and record the distribution of the outcomes to provide further insights into the behavior of the system.

In this chapter, we will discuss the methods used to create stochastic models and how these are implemented in both MATLAB® and Python. Using simple examples, we will illustrate how the models can be run and interpreted. Some examples of stochastic models from a variety of fields will be presented. Finally, we provide some exercises where you build a stochastic model.

11.2 CREATING A STOCHASTIC MODEL

The initial steps for creating a stochastic model are the same as those presented in Chapter 1. The model objectives are specified leading to a selection of the variables and the governing equations. For a stochastic model, the first additional step is to select the variables whose values will vary randomly. For each of those variables, several additional decisions need to be made:

- Associate a probability distribution with the variable that reflects the nature of the random variation of the variable. This could come from experimental data, from other observations of the system, or from the design parameters of an engineered system.

- Create a relationship between any given value or range of possible random numbers and the value of the variable that will be inserted into the simulation run.

- Choose an appropriate random number distribution from which to draw the numbers for the simulation. Most of the random number generators choose real numbers between 0.0 and 1.0. If we want all values of the random sequence to have an equal chance of occurring, we will use a uniform random number scheme. On the other hand, it may be that we expect the distribution to be normal (a Gaussian or bell-shaped curve) where it is more likely to get values closer to the mean and less likely to get values at the either end of the distribution. We will discuss this further in Section 11.3.

- Decide how many times we will run the model with the different random variables.

- Prepare the appropriate code to draw random numbers, run the model, and store the results from each run.

- Analyze the distribution of all the runs to gain insights into the system being modeled.

A simple example should help to illustrate these steps. Recall that in the traffic exercise, we assumed that traversing an intersection with a stop sign would take 30 seconds on average. That would include the time to decelerate, stop, and accelerate back to the maximum travel speed on that street segment. However, we know from experience that it will take less time if there are no conflicts with other cars or pedestrians at the intersection and more time if such conflicts occur.

We can add a random variable to our model that varies the amount of time at each stop sign depending upon the number and nature of conflicts that occur at that intersection. To implement this, we could make observations of the target intersections along our route during the time of day we would normally commute to work. Let us say we divided the events of cars entering these intersections to the four categories in Table 11.1. For those observations, we record the time at the intersection and the number of times that event occurred. The table is then compiled showing us the

TABLE 11.1   Intersection Conflict Time Delays and Probabilities

Intersection Event	Time Delay (seconds)	Percentage of Events
Stop sign with no conflicts	10	30
Stop sign with single car conflict	20	20
Stop sign with pedestrian conflict	30	20
Stop sign with more than one conflict	60	30

range of time from 10 to 60 seconds along with the percentage of the time that each event occurred. We can then use that distribution to alter our program to include these random events.

One way to represent this distribution in our revised code is to draw a uniform random number between 0.0 and 1.0. We could then insert branching statements assigning an appropriate value of the time taken based on the result of the random variable. For numbers between 0.0 and 0.30, we would assign the value 10 seconds, for those between 0.31 and 0.5, 20 seconds, for 0.51–0.70, 30 seconds, and for 0.71–1.0, 60 seconds. We would then add a loop to run the model a large number of times—say 100—and then evaluate the distribution of times and their impact on our commute time along these routes.

If we wanted to add additional reality to our model, we could create another random variable to choose weather conditions based on the historical weather records and assign slower speeds to travel on days with fog, rain, or snow.

There are other ways in which random variation can be used in models in addition to using a random number generator for inserting random effects into those models. Where there is a large dataset representing a particular input or outcome, one can draw a random sample from the dataset to examine the potential range of outcomes that could occur. For example, one could draw a random sample of weather events during a particular season to insert into models of the hydrological response of a stream. Similarly, there may be a large dataset representing people's home purchasing behavior under different market conditions that could be applied to a model of the housing market.

Another way that stochastic processes can be used is in the estimation of complex mathematical integration. Stochastic models are used to estimate the area under a curve by counting the proportion of points that fall within finite subareas of the distribution.

11.3 RANDOM NUMBER GENERATORS IN MATLAB® AND PYTHON

Both MATLAB and Python have several options for generating pseudorandom numbers. The major options are shown in Table 11.2. One important characteristic of all pseudorandom number generator algorithms is that they start with a *seed* from which all subsequent numbers are drawn. If you start the algorithm with the same seed, you will get the same sequence of random numbers. Those are not truly random.

TABLE 11.2   Random Number Generators in MATLAB® and Python

Command	Description
MATLAB Commands	
rand	Uniformly distributed random numbers
randn	Normally distributed random numbers
randi	Uniformly distributed pseudorandom integers
rng	Control random number generation
random(pd)	Returns random number from a probability distribution
Python Commands	
Random module (import random) rand	Subcommands to generate uniform, normal, or other distributions of random numbers
random.random()	Uniform random numbers
random.gauss(mu, sigma)	Normal random numbers with mean mu and standard deviation sigma
numpy.random	Module with many similar random number generators

However, that characteristic is sometimes useful when debugging a program to see if the answers match. To get closer to truly random numbers, the seed should be varied.

Table 11.2 shows that MATLAB has three different random number generators for uniform, normal, and integer random numbers (MathWorks, 2016). Each time MATLAB starts, it resets to the same initial seed for random numbers. The rng command allows control over the seed as well as the algorithm that is used in generating random numbers. The command can be used to save a particular starting point, reset to a starting point, set a user defined seed, and set the generator algorithm. There is also a function that can draw a sample for any defined probability distribution.

In Python, there are at least two sets of random number functions. One resides in the random module and has separate commands for a variety of distributions (Python Documentation, 2016). If no seed is provided using a random command, the default is to set the seed to the system time. NumPy also has a variety of random number and sampling commands that operate in a similar way (NumPy Documentation, 2016).

11.4 A SIMPLE CODE EXAMPLE

Let us create a simple stochastic model to represent the random toss of a coin. We can select a uniform random number between 0.0 and 1.0. Assuming that the coin is balanced, there should be an equal probability of getting a head or tail on each toss. Thus, if the random number that is generated is less than 0.5, we can say that it simulated heads and if it is 0.5 or greater, it simulated tails. The following MATLAB and Python codes show one approach to implementing this program:

MATLAB CODE FOR COIN TOSS SIMULATION

```
r=zeros([100:1]);
heads=0;
tails=0;
rng('default')
for j=1:100
    r(j)=rand();
    if r(j) < 0.5
        heads=heads+1
    else
        tails=tails+1
    end
end
heads
tails
```

PYTHON CODE FOR COIN TOSS SIMULATION

```
import math
import numpy as np
import random
r=np.zeros(100)
heads = 0
tails = 0
random.seed(8952)
for j in range(100):
    r[j]=random.random()
    if r[j] < 0.5:
        heads = heads+1
    else:
        tails = tails+1
print("heads= ",heads,"tails=",tails)
```

In each code, we start by creating a vector of 100 to hold a set of random numbers. We then initialize the random number generator. In MATLAB, we can start with the default random number seed that occurs when the program starts. Alternatively, we could insert our own numerical seed. In Python, we insert our own seed. The default in Python is to use the system time as a seed. We then have a loop where we draw a random number and increment either the head or tail variable depending on the outcome. Finally, we print the values of the final count. Try out the code and see what happens.

In both cases, we have provided a constant seed. Thus, if you run the program multiple times, you should get the same answer since the same pseudorandom numbers will be generated. This is an approach we might use while debugging a program to make sure that there is nothing else that impacts the outcome in unexpected ways. If we then want to run it with a truly random sequence, we can change the code to use a seed, which changes on each run by using something like the date and time or another random number. We would then need to add code to save the final outcomes of each run and then examine the distribution of the outcomes to gain insights into the system behavior. We might also choose to increase the number of iterations with the loop in a way, which better approximates the number of possible occurrences of the phenomenon under study per unit time.

11.5 EXAMPLES OF LARGER SCALE STOCHASTIC MODELS

There are many circumstances in which a Monte Carlo modeling approach may be used. Those include the potential failure of components in complex engineered systems, economic models of buying behavior in a market, the movement of molecules in a solution (molecular dynamics), the impact of travel behavior on roadway congestion, and changes in metropolitan land use over time. We have chosen three examples to describe in more detail here to illustrate the range of problems that can be addressed within this modeling framework.

Kalos (1970) provides an early example of the use of Monte Carlo modeling of the neutrons in a nuclear reactor. Neutrons are emitted by fissionable materials to instigate a chain reaction while other materials in the reactor serve to slow down the neutrons from fission, and neutron absorbing materials control the reactor to protect the area outside the reactor. Each of the steps in the life of a neutron is in part a random process. This includes the fission that creates the neutron, the direction it moves after it is created, and the speed of that movement. The neutron moves in a straight line at a speed that is governed by a probability equation. It may then collide with other atoms or with other materials in the reactor impacting the outcome of each movement. Those include causing a new cycle of fission, scattering in a new direction, or disappearing because it is absorbed. A fully functional model of all of these reactions requires the tracking of a very large number of neutrons through many random cycles with different arrangements of the fuel rods, pressure vessels, coolant pipes, and other reactor characteristics. Bareiss (1970) provides an indication of the computational power required for different components of reactor simulation.

Another situation in which Monte Carlo modeling has been used is in managing service levels in retail establishments. For this example, let us assume that we are the manager of a bank managing the number of tellers that are available on various days of the week and times of the day. Ideally, we want enough tellers on hand so that the wait time for customers is acceptable while not having so many tellers that they have large chunks of idle time. To solve this problem, we need to have a sufficient sample of actual customer demand over the periods we are simulating. Those data can then be converted into a probability distribution from which we can generate a random set of customers arriving at the bank for service. In addition, we need to know the distribution of the amount of time it takes for the teller to complete the transactions for a single customer. We would expect a range of transaction times from a simple deposit or check cashing to a more complex combination of multiple deposits, withdrawals, and/or other services. We could then simulate the random arrival of customers approximating their distribution from the sample data, the transaction times, and the resulting wait and idle times with different numbers of tellers available. In this way, we could arrive at a schedule for the tellers, which provides an acceptable waiting time for customers and low idle times for the tellers. This approach could be applied to any retail service with the potential for queuing customers awaiting service. A similar framework has been used to simulate the impacts of traffic on congestion, the routing of delivery services, and many other situations where queuing impacts the efficiency of service delivery.

Keane et al. (2011) have created a comprehensive model for exploring fire and vegetation dynamics. The model is a spatially explicit representation of the forest stand and the impacts of climate conditions on the probability and extent of widespread forest fires. The model defines polygons of forest cover and then specifies a number of site parameters related to the nature of the forest stand and ecosystem features. Both deterministic and stochastic components of the model are used to simulate the potential for fire and its progression under a variety of conditions. Stochastic components include the influence of seed trees in seed dispersal, the potential for a fire to spread from one area to others, and the long-term climatic pattern.

A complex set of tree stand level and landscape-level interactions are simulated taking into account a number of different biophysical processes, the growth of the tree stands and undergrowth, and weather conditions. These are then used to project the impacts of these interactions on the potential for fire and its spread in a particular landscape.

Another spatially explicit landscape model by Voinov et al. (2004) was used to simulate land use changes in the Patuxent River Basin in Maryland and the impacts of those changes on water runoff and water quality. The economic land use conversion model embedded in this effort used data on property sales, the economic and ecological characteristics to estimate the probability of the conversion of land from agriculture or forest into different densities of urban development. The relative likelihood of land conversion was estimated using an empirical analysis of the historical changes in land use in the region based on factors such as distance to employment centers, access to infrastructure, and proximity to other desirable and undesirable land uses. The land use conversion results were then passed to simulation routines that estimated water runoff and water quality and its potential impacts on the health of the watershed.

Wilkinson (2011) provides an introduction to the need for stochastic modeling in biology. He notes that even a simple model of bacterial population growth needs to account for the discontinuous concentration of bacteria and the stochastic nature of their growth. He notes that most biochemical processes are driven by Brownian motion where a discrete number of biomolecules interact when they meet at random times. He provides examples of how stochastic modeling provides insights into biological processes.

Lecca et al. (2013) provide an excellent description and comparison of the deterministic and stochastic approaches to modeling biochemical processes. They go on to describe several of the major algorithms used to model the stochastic processes in biochemistry. They also describe how stochastic modeling is applied in systems biology. They use a specialized programming language developed in Italy called BlenX that can be used to address some of the challenges with modeling complex biological systems.

Many large-scale models include both deterministic and stochastic components. Historically, simpler deterministic models have often added stochastic components to explore the impacts related to parameters that vary over time or space. As modeling efforts have evolved over time, many deterministic models have added stochastic components to provide a more realistic picture of the range of possible outcomes.

EXERCISES

1. Create a model of a pair of dice analogous to the coin model where there is an equal probability for each of the six numbers on each die. Run the model 100 times and make a graph of the resulting distribution. Create a second version where the dice are biased so that there is a 30% greater chance of a six on each die. Compare the two graphs and describe the results.

2. Assume that there is a hiker who is lost in the forest and is trying to find their way out. A hiker without a compass trying to find their way in the dark can step in any of eight directions (N, NE, E, SE, S, SW, W, NW) with each step. Studies show that people tend to veer to the right under such circumstances. Initially, the hiker is facing north. Suppose at each step, probabilities of going in the indicated directions are as follows: N—19%, NE—24%, E—17%, SE—10%, S—2%, SW—3%, W—10%, NW—15%. Notice that these probabilities add up to 100%. We are going to construct a random walk simulator that uses the probability and the built-in random number generator in MATLAB and Python. You know that the random number generator provides a uniform distribution of numbers over the range from 0 to 1. This is what will be used in our simulator. For MATLAB users, download the MATLAB script, hikerwalk.m, and two function files Hiker.m and animWalk.m. For Python users, download the file hikerwalk.py. These are an implementation of the model we have been discussing. The user provides the number of steps, and the script will return an animation of the hiker's random walk, the coordinates of the hiker's final position, and the hiker's distance from the origin. Run the hikerwalk script using 100 steps. The resulting random walk will be animated in a figure window, which is scaled to contain the entire walk.

Let us conduct a number of random walks to identify a search grid for the lost hiker. Using the script, conduct 20 random walks. Record the coordinates of the hiker's final position and distance from the origin. Plot those results and use them to recommend a search area for the lost hiker.

FURTHER READINGS

Python Software Foundation. Python random.py. https://docs.python.org/3/library/random.html (accessed December 1, 2016).

SciPy.org. Random sampling. Numpy random. https://docs.scipy.org/doc/numpy/reference/routines.random.html (accessed December 1, 2016).

REFERENCES

Bareiss, E. H. 1970. Computers and reactor design. In *Computers and their Role in the Physical Sciences*, ed. S. Fernbach and A. Taub. pp. 337–384. New York: Gordon and Breach Science Publishers.

Kalos, M. H. 1970. Monte Carlo methods. In *Computers and their Role in the Physical Sciences*, ed. S. Fernbach and A. Taub. pp. 227–235. New York: Gordon and Breach Science Publishers.

Keane, R. E., R. A. Loehman, and L. M. Holsinger. 2011. The FireBGCv2 landscape fire and succession model: A research simulation platform for exploring fire and vegetation dynamics. General Technical Report RMRS-GTR-255. Fort Collins, CO: U.S. Department of Agriculture, Forest Service, Rocky Mountain Research Station.

Lecca, P., I. Laurenzi, and J. Ferenc. 2013. *Deterministic Versus Stochastic Modelling in Biochemistry and Systems Biology*, Woodhead Publishing Series in Biomedicine. Jordon Hill, Great Britain: Woodhead Publishing. ProQuest ebrary. Web. 8 (accessed December 2016).

MathWorks. Random number generation. https://www.mathworks.com/help/matlab/random-number-generation.html?requestedDomain=www.mathworks.com (accessed December 1, 2016).

Voinov, A. R., R. Soctanza, M. J. Roelof, T. M. Baoumans, and H. Voinov, 2004. Patuxent landscape model: Integrated modeling of a watershed. In *Landscape Simulation Modeling*, ed. R. Costanza and A. Voinov, 197–232. New York: Springer Verlag.

Wilkinson, D. J. 2011. *Stochastic Modelling for Systems Biology*, Chapman and Hall/CRC Mathematical and Computational Biology 2nd ed. London: CRC Press. ProQuest ebrary. Web. 8 (accessed December 2016).

CHAPTER 12

Functions

Although we have been writing scripts to allow us to reexecute code, programming languages like MATLAB® and Python allow us to create reusable blocks of code that can accept parameters to execute against *functions*. We have already been using built-in functions, such as various math libraries or bits of administrative code around array and matrix creation, but we can create our own functions, and use them just like we have been using the built-in functions.

Functions provide modularity and allow easier code reuse and these are a very powerful programming tool. Object-oriented programming is also supported by both MATLAB and Python, and it is a very useful technique, but it is beyond the scope of this book.

12.1 MATLAB® FUNCTIONS

We will introduce various capabilities of MATLAB's support for user-defined functions by building an example function and adding in additional complexity. First, we will create a function called *stats* by creating a file called stats.m in our working directory, containing the following code:

```
function mean = stats(x)
if ~isvector(x)
        error('Input must be a vector')
end
mean = sum(x)/length(x);
end
```

This function will take the input argument and place it in the variable "x," check that x is a vector (and return with an error if it is not), and then calculate the mean for the vector, store that in the variable *mean*, and then return it. It is important to note that MATLAB handles input variables by something commonly called *pass by value*, which means that any modifications to the input variable that occur inside the function will not be reflected to the variable outside the function. If you saved the file in your MATLAB path (which includes the current working directory), you can test this by calling the function in the Command Window, passing different values and variables to it.

MATLAB functions also support multiple return values. We can modify the function in stats.m to return the mean and standard deviation of the input variable.

```
function [mean, sdev] = stats(x)
if ~isvector(x)
        error('Input must be a vector')
end
len = length(x);
mean = sum(x)/len;
sdev = sqrt(sum((x-mean).^2/len));
end
```

Now, when you call the function, you can get both the average and standard deviation returned and stored in local variables, by calling your function with the following syntax:

```
[avg, stdev] = stats(input_vector);
```

MATLAB functions are in scope and callable provided the file name matches the function name, and the file is located in the MATLAB path. The path can be modified by clicking *Set Path* in the environment section of the Home ribbon, and can be inspected by looking at the *path* variable. You can create what are called subfunctions—functions only available and in-scope inside a particular function—by simply adding additional *function* definitions inside your primary function. This can be useful if you have code you want to reuse multiple times in a function, but that has no value outside of that function.

MATLAB also supports variable numbers of input and output arguments by using special variables called *varargin* and *varargout*. These can be used to create functions that have optional input or output variables. This will not be commonly used, so we will not talk about it in any more detail here.

Variables defined inside of a function only have scope inside that function, meaning that when the function exits, the variables cannot be accessed, and if they have a name that conflicts with a name in the calling workspace, the variable in the calling workspace is not affected.

12.2 PYTHON FUNCTIONS

Python allows the creation of local functions, which are only in scope within the currently executed file, and the creation of code libraries via modules. We will first look at the syntax around functions and then proceed to demonstrating how to create custom modules.

12.2.1 Functions Syntax in Python

Python functions have very simple syntax. The *def* keyword is used to define the function and input variable list, and the *return* keyword is used to mark the function end and return any calculated value(s). We can demonstrate this simplicity with an example that calculates the mean for an input vector. You can place this code into the Editor window in Spyder to test it.

```python
import numpy as np

# Function definition is here
def stats(my_array):
  "This calculates the mean of the input array"
  mean = np.sum(my_array)/np.size(my_array)
  return mean

# Now you can call stats function
m = stats([4,5,6])
print(m)
```

As we have not introduced modules yet, you will note that we must define and use the function in the same Python file. It is important to note that Python handles input variables via a convention commonly called *pass by reference*. This means that modifications to the variables passed to a function made inside the function are visible outside of the function. For example, if you were to change a value at any position inside "my_array" in our example function above, that change would be visible in the variable "my_array" as it was used after the function was called.

If you call stats() without an input argument, it will generate an error as "my_array" is a required argument. Python allows us to specify input variables several ways and call a function a different way to create more flexibility in input arguments. The alternate way of calling a function is to use the input variable name in the function call. This lets the interpreter determine what input variables map to which arguments, and allows calling the function with the input arguments in any order. For example, you can call *stats* like this:

```
stats(my_array=[2,3,4])
```

You can also provide default values for any or all input arguments, effectively turning those arguments into optional values. If we wanted to specify a default value for my_array in the stats() function, we could modify the function definition like this:

```
def stats(my_array = [1, 2, 3]):
```

Now, when we call the stats() function, if we did not supply an argument for my_array, the Python interpreter will use the array [1,2,3] instead of an error stopping execution of our program.

Python also supports variable-length argument lists. These additional arguments are nonnamed, cannot have default values, and are placed in a *tuple*, a special type of list in Python that you can iterate over.

Variables inside a function in Python have local scope, which means that they can only be accessed inside the function, and have no value outside of the function. If a variable name inside a function is the same as a variable that exists outside the function, the variable outside the function is not impacted by anything that happens inside the function.

12.2.2 Python Modules

Code reusability is one of the biggest advantages of functions, and if we could only use functions in the file they are defined in, our code would be less reusable and become more difficult to maintain. Thankfully, Python allows us to create our own user-defined modules to package up our custom functions to make them easier to reuse.

Creating a module is very simple. If we want to package up our stats() function into a module called *mymodule*, we would place the function definition (along with any other functions we wish to put into *mymodule*) inside mymodule.py. At this point, we can import the module and use the functions defined in it just as we have imported other modules such as NumPy.

```
import mymodule
a = mymodule.stats([1,2,3,4,5])
```

When the Python interpreter attempts to import modules, it first looks in the local working directory and then looks in the environment variable PYTHONPATH for a list of directories to search.

EXERCISES

1. Write a function that calculates the factorial for the input value.

2. Write a function that calculates the factorial for each value of an input vector and that returns a vector of results.

3. Write a function that takes an input matrix and calculates the average value of each row (or—if instructed by an optional argument—the average value of each column) and that returns a vector containing the results.

4. Package your functions from these exercises into a custom Python module.

CHAPTER 13

Verification, Validation, and Errors

13.1 INTRODUCTION

Knowing whether or not a model is a reasonable representation of the system being studied is a critical question for all modeling efforts. This turns out to be a very difficult question to answer. Since every model is a simplification of reality, all models will deviate from 100% accuracy when compared with available experimental data. At the same time, experimental data may contain errors, be available for only limited cases, or be lacking in its scope in other ways, making it impossible to fully prove that a model is *correct*.

The modeling process itself is dynamic. We often start out with a very simple representation of a system. The results of that effort are then evaluated with respect to the available knowledge of the system behavior and the insights provided with the initial model. With this analysis in mind, we often change the model design by adding additional components, altering the mathematical representation, and changing the computer algorithms used. We then need to judge whether the newer versions of the model are better representations of the system. All such efforts are constrained by the resources that we have in place to undertake the modeling effort—the scientists, programmers, data, and computational resources required to make the improvements.

Judging how well we are doing with our modeling effort requires a framework that we can use to judge our results. The processes that have been developed to make those judgments are called verification and validation. Verification is the process used to confirm that a model is correctly implemented with respect to the conceptual model it is based on. Validation is a check of the representation of the model to reality. Several professional organizations have sanctioned their own definitions of these terms and the processes that are used to achieve their goals. We will discuss those later in this chapter.

Section 13.2 provides a review of the sources of errors and how we measure them. We then review the definitions and scope of the verification and validation process in more depth. The methods used to undertake these processes are then summarized followed by a few exercises.

13.2 ERRORS

There are many possible sources of error in scientific research and modeling. The data that are used to support the development of models may have errors. Those errors can be associated with the accuracy of the instruments that are used to measure what is happening. Errors may also be made in the formulation of a model by making incorrect assumptions about the system behavior, assumptions that oversimplify the system, decisions on the critical components of the model, and the variability of model parameters.

The implementation of a model may also produce errors. Those can include programming and logical errors that incorrectly implement the model representation, calculation errors associated with the representation of numbers by the computer, and errors in the algorithms used to approximate the mathematical basis of the model.

Identifying and separating out the many sources of errors is a challenge. The key to approaching this problem is a deeper understanding of the sources of errors and the techniques that can be used to minimize their impact on the model results.

13.2.1 Absolute and Relative Error

For instances where we know the real or experimental value of a quantity, we can calculate the difference from the modeled value in two ways. Of course, this assumes that there are no measurement errors associated with the experimental value we are using for comparison.

The absolute error is the absolute value difference between the model result value $\bar{X}$ and the assumed correct value of X:

$$\left| X - \bar{X} \right| \tag{13.1}$$

The relative error is the absolute error divided by the correct value, assuming the correct value is not zero. This can be calculated as a proportion or a percentage:

$$\text{Relative error} = \frac{(X - \bar{X})}{X} \tag{13.2}$$

or

$$\text{Relative error} = \frac{(X - \bar{X})}{X} * 100\% \tag{13.3}$$

We will use these measures to illustrate the nature of computational errors that occur in modeling. Whether a particular level of error is acceptable depends upon the objective of our model and the risks.

13.2.2 Precision

Decimal numbers in a computer are represented as binary floating-point numbers. Those numbers are represented in the computer using scientific notation that includes a sign, a mantissa or significand representing the numbers and an exponent, for example, $+8.94357 \times 10^2$ to represent 894.357. The number of significant digits in the mantissa is limited by the amount of memory used in the computer to store that number. Numbers that are represented in 32 bits are called single precision numbers. They can have six to nine decimal digits. The range of numbers that can be represented in single precision is from 10^{38} to 10^{-38}. Double precision numbers can store up to 15–17 decimal digits and represent numbers from 10^{308} to 10^{-308}.

13.2.3 Truncation and Rounding Error

The default floating-point representation in MATLAB® and Python is double precision. For the models we are running for this course, there are no constrictions on available memory so everything can easily be done in double precision. However, it is important to understand the potential errors that arise from the differences in precision (Bush, 1996).

The first possible problem is called truncation error. This will most frequently occur when using very large or very small numbers. With fewer digits in the mantissa, values are truncated and thus take a slightly different value. Even if this value is small, it may accumulate if it is used in a loop where it adds an incremental error on each iteration. When we make a division that results in an infinite decimal expansion, even the double precision number will truncate the value.

The second related problem is rounding error. This may occur when we add a very large and very small number. Given the conversion of the decimal numbers to binary, the small number may be ignored in the resulting total. Rounding also occurs when using the default formatting of many programming languages. The default may present a decimal number rounded to four or five significant digits, rounding off the number based on the next significant digit in the sequence.

To illustrate these problems, download and run the program truncation.m or truncation.py. The program codes are shown as follows:

TRUNCATION CODES

MATLAB Code
```
%Truncation example
clear all;
format long
x=3.56e6;
y=2.2800e6;
a=single(x);
b=single(y);
z=x*y;
c=a*b;
fprintf('\nz = %2.8e c= %2.8e',z, c)
format short
z
c
```

Python Code
```
# -*- coding: utf-8 -*-
"""
Created on Thu Dec 15 13:52:27 2016
Truncation example
"""
import numpy as np
x=3.56e15
y=2.2800e22
a=np.float32(x)
b=np.float32(y)
z=x*y
c=a*b
print ("z=: %2.8e, c=: %2.8e" % (z, c))
print (z, c)
```

In these codes, two very large numbers are input as x and y. A second version is created in single precision as variables a and b. Both pairs are then multiplied. The result is shown in a format, which shows the full value of each. Although the numbers are close, the single precision number is slightly smaller. In both cases, the program writes out both numbers in a more standard decimal format. When this is done, rounding occurs and both values appear to be the same.

The conversion between decimal and floating-point numbers often results in these errors. This is an important tendency to keep in mind when writing a program using floating-point arithmetic. It implies that we should never test for the equality of two floating-point numbers because there is high probability that they will be close but not equal.

One way of minimizing errors associated with rounding and truncation is to perform the calculations with very small numbers first and then do the calculations with larger numbers.

13.2.4 Violating Numeric Associative and Distributive Properties

Computer arithmetic can violate the associative properties of addition, subtraction, and multiplication due to rounding and truncation errors and the resulting representation of decimal numbers in binary formats. Normally, we would expect the following to be true:

$$(a + b) + c = a + (b + c) \quad \text{and} \quad (ab)c = a(bc) \tag{13.4}$$

Rounding and truncation errors may produce errors that result in slightly different results when the order of computations is changed. This will be illustrated by one of the exercises.

13.2.5 Algorithms and Errors

Numerical errors have been extensively evaluated as computer simulation has become such a large part of science and engineering practice. As various mathematical representations have been used in simulations, algorithms have been designed and tested to provide the most efficient computational requirement and the procedure used on the computer to perform a calculation. Those algorithms have become parts of scientific libraries that can be inserted into codes as function calls. This avoids the need to write a program from scratch that performs the same calculation and avoids the errors that might arise from such a code.

In MATLAB, many of these algorithms are available as part of the various toolboxes that come with the program. In Python, they are parts of the mathematic and scientific libraries. In both cases, there are specialized libraries that focus on particular fields of study for example tools for bioinformatics, astrophysics, image processing, and statistical analysis (see Mathworks, 2016b; Wikipedia, 2016). Tracking the available modules and packages in Python is a challenge as new packages and routines are constantly under development by the user community. As a rule, those libraries should be consulted before writing new code to represent a particular system or mathematical operation.

13.2.5.1 Euler's Method

We will use our approach to solving an ordinary differential equation to illustrate how the algorithm being used can impact our computational results. For several of our previous assignments, we have solved an ordinary differential equation by changing it to a difference equation and calculating incremental changes in the quantity with respect to time. Recall the models for unconstrained population growth and the velocity of a ball tossed off a bridge. Our estimate for the value at time t was based on its value at time $t-1$ plus the change that occurred in that period of time. That algorithm is more formally known as Euler's method after the eighteenth century Swiss mathematician Leonhard Euler.

The method can be applied to discover the shape of the curve described by an ordinary differential equation given the starting point of the curve and solving for the slope of the tangent line to the curve at each subsequent point. More formally the equation is

$$y'(t) = f(t, y(t)) \tag{13.5}$$

where:

$y'(t)$ is the estimate of the value of y given the time dependent function
  $f(t, y(t))$
$y(t_0) = y_0$ is a known origin

To implement this algorithm, we choose a step size h and then step through the calculation:

$$y_{n+1} = y_n + hf(t_n y_n) \tag{13.6}$$

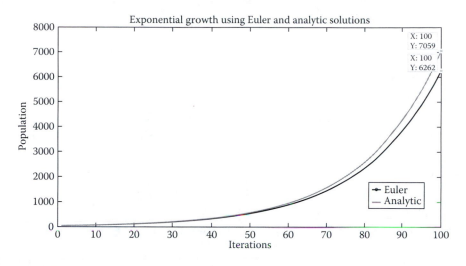

FIGURE 13.1   Graph of Euler versus analytic solution.

Let us illustrate the results using an exponential population growth example. Let us say we have a starting population of 50 and a growth rate of 0.05 annually. Our equation then is

$$\frac{d_P}{d_t} = rP \tag{13.7}$$

where:
  $P_0 = 50$
  $r = 0.05$

The analytical solution to this equation is the exponential growth equation:

$$P = 50e^{0.05t} \tag{13.8}$$

Figure 13.1 shows the result of the simulation after 100 iterations using both the Euler method and the analytical solution with the time step set to 1. The figure shows that the Euler method creates a pretty accurate approximation until iteration 50 when it starts to diverge from the analytical results. By the time it reaches the 100th iteration, the difference is 797 or 12.7% error. We can reduce the error by using a smaller time step of 0.25 at the cost of four times the computational time. When this is done, the error is reduced to about 3.8%. For a small program such as this one, that computational burden is not significant but for any large code, the trade-off between accuracy and computational requirements can make a huge difference in the utility and feasibility of the code.

13.2.5.2 Runge–Kutta Method

Carl Runge and Martin Kutta were both German mathematicians who derived a four-step method for solving ordinary differential equations around 1900, well before the algorithm was implemented in computer code. The method has thus been named Runge–Kutta 4 or RK4. Mathworks (2016a) provides a nice set of overview videos on this and other ordinary differential equation (ODE) solution algorithms with examples from MATLAB.

Given the same basic problem as defined by Equation 12.5, the algorithm goes through four steps to arrive at this solution:

$$y_{n+1} = y_n + \frac{h}{6}(k_1 + 2k_2 + 2k_3 + k_4) \tag{13.9}$$

For $n = (0, 1, 2, 3, \ldots$ no. steps).

$$k_1 = f(t_n, y_n) \tag{13.10}$$

where k_1 is the slope at the beginning of the interval using Euler's method.

In the second step, a second increment is calculated based on the slope of the midpoint of the interval:

$$k_2 = f\left(t_n + \frac{h}{2}, y_n + \frac{h}{2}k_1\right) \tag{13.11}$$

The third increment k_3 is also based on the midpoint but uses the value of k_2:

$$k_3 = f\left(t_n + \frac{h}{2}, y_n + \frac{h}{2}k_2\right) \tag{13.12}$$

Then the fourth increment is based on the slope at the end of the interval using $y + hk_3$:

$$k_4 = f(t_n + h, y_n + hk_3) \tag{13.13}$$

The final estimate as shown in Equation 13.9 is a weighted average of the four coefficients with the middle two coefficients given greater weight. For our simple example, you may undertake the exercise that

TABLE 13.1   Selected ODE Solvers in MATLAB® and Python

Function	Description
MATLAB Functions[a]	
ode45	Listed as the first solver you should try
ode23	Listed as more efficient at problems with low tolerances
ode15s	Used when ode45 fails and on problems with stringent error tolerances or when solving differential algebraic equations
Python Functions	
scipy.integrate.odeint	Integrate a system of ordinary differential equations
scipy.integrate.ode	A generic interface class to numeric integrators
scikits.odes 2.2.1	Package that installs several other ODE solvers for Python

[a] Consult the MATLAB and Python documentation for other options and for solver syntax.

shows that the RK4 solution to the equation matches the values of the analytic solution after 100 iterations.

> Choose the most appropriate algorithm for calculating any specific mathematical function. Look up the function in the programming environment help documents to find the choices of algorithm for any specific problem.

13.2.6 ODE Modules in MATLAB® and Python

There are several ODE modules in MATLAB and Python that can be used to solve ordinary differential equations of different types. These are shown in Table 13.1. Both of the programming environments have multiple solvers that can be applied to ODEs with different characteristics. For the problems involved in this book, the first example in each—ode45 in MATLAB and scipy.integrate.odeint—should suffice for the problems and project we undertake.

13.3 VERIFICATION AND VALIDATION

Given the previous discussion of the many sources of errors and the difficulty in measuring those errors, it should be no surprise that model verification and validation (V&V) is a challenge. In this section, we provide some historical background to the process of V&V and the formal frameworks that have been established by several professional organizations. We then go on to provide some practical guidelines for conducting V&V that you can immediately implement.

13.3.1 History and Definitions

The Society for Simulation was one of the first organizations to create a set of formal terms relating to V&V (Schlesinger, 1979). They divided the modeling process into three major elements—reality, a conceptual model, and a computerized model as illustrated in Figure 13.2. The conceptual model is the abstraction of reality that is used as the basis to create a computerized model including the governing relationships and equations that are attempting to describe reality. The conceptual model also defines a domain of intended application, which they define as the "prescribed conditions for which the conceptual model is intended to match reality."

This leads to the computerized model through a programming effort. They then defined model verification as "substantiation that a computerized model represents a conceptual model within specified limits of accuracy." As we will see this early definition has carried forward to some of the current standard frameworks for verification.

Going on in the modeling process, the computerized model is compared to reality by using computer simulations to try to describe that reality. Their definition of model validation then becomes "substantiation that a computerized model within its domain of applicability possesses a satisfactory range of accuracy consistent with the intended application of the model." How we judge whether a model sufficiently accurate depends on its intended use. For example, we might have a model of a rocket launch that places an object in orbit around the earth. If our purpose is to put it

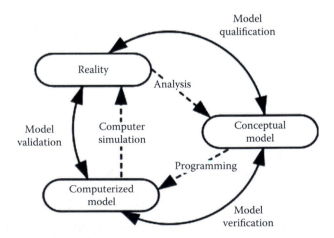

FIGURE 13.2 Development of concepts and terminology for modeling and simulation.

into a stable orbit at a particular altitude within a few hundred meters and we can validate that our model will do that based on previous launches, the model is valid. However, if that object must rendezvous with the space station, the required level of accuracy is within a few meters and our more general model would not be valid. Similarly, we might undertake a regional population forecast to provide general insights about the growth of the region and the possible impacts on the economic base and social change. However, if we are going to invest in a new water system including water lines and a water treatment facility, investing millions of dollars in facilities, we need a much more accurate and spatially explicit model of the growth so that we know where and when to start building new facilities.

These early definitions of V&V helped to guide several other professional organizations to create standards for the process. Oberkampf and Roy (2010) describe the history of V&V with a number of examples of frameworks adopted by various professional groups. A set of definitions created by the Department of Defense community in 1994 was very similar in scope to that of the Society for Simulation. That definition has been used with only minor changes to the present day and is shown as follows:

Verification: The process of determining that a model implementation accurately represents the developer's conceptual description of the model.

Validation: The process of determining the degree to which a model or simulation and its associated data are an accurate representation of the real world from the perspective of the intended uses of the model.

Source: Department of Defense, DoD modeling and simulation (M&S) verification, validation, and accreditation (VV&A) Department of Defense Instruction Number 5000.62. https://www.msco.mil/vva.html, 2009.

The American Society of Mechanical Engineers has added several explicit steps to the verification process (ASME, 2006). In particular, they specify two verification steps—code verification and calculation verification. Code verification is defined as "the process of determining that the numerical algorithms are correctly implemented in the computer code and of identifying errors in the software." Calculation verification examines the correctness of the input data, the numerical accuracy of the solution, and an evaluation of the correctness of the output for the simulation runs. The correctness of the input data involves an assessment of possible human error in using the wrong datasets or parameters for model runs.

This is one of the most difficult types of errors to find, especially with large-scale models with many inputs. The numerical accuracy in this case is associated with the algorithms used to translate the mathematics into the computer code and not the accuracy with respect to the model validation. The checking of the output is aimed at evaluating the quantity and nature of the output distributions to ascertain that they produce the expected range and distribution of values. For example, an anomaly in the shape of an output curve may indicate that the code is incorrectly calculating a subset of the data. Values that exceed the physical limits of a variable or that are extremely large or extremely small may indicate that the code is misrepresenting the system in some way.

The definition of validation focuses on comparisons of the model results with the real system it represents relative to the intended use of the model. It is important to note that the level of accuracy associated with validation is not absolute but is dependent upon the use for which the model was developed. Regardless, validation can involve both quantitative and qualitative assessments of the models behavior with respect to a real system. The nature of those measurements depend upon the availability and quality of data relating to the modeled system. Where possible it should involve an assessment of validity of the parameters used as inputs to the model as well as comparisons with one or more of the outputs.

13.3.2 Verification Guidelines

There are a number of formal frameworks that can be applied to complete the verification of a model. Those include standard software engineering best practices from the computer science community, code verification criteria from the scientific computing community, and specific tests associated with different classes of codes such as finite element methods. Those detailed approaches are summarized by Oberkampf and Roy (2010) and Murray-Smith (2015) and are beyond the scope of our introduction to modeling. We have extracted a subset of these approaches to provide a checklist of practical steps that should be taken to verify even simple models. Those guidelines are shown in Table 13.2.

Since the development of a simulation model is dynamic, the guidelines in Table 13.2 should be followed as each section of code is developed. This approach will avoid many problems of trying to find a bug in the much larger completed code. As each code segment is developed, inspect and debug that segment to make sure that the proper data and calculations are passed to the next code segment. Once the code is completed,

TABLE 13.2   Checklist of Verification Steps

Action	Description
Choice of algorithm	Ensure that the algorithms chosen for the model are the most appropriate for the circumstance
Software inspection	Careful review of each program module to ensure that the algorithms are properly programmed, parameters are properly input and used, and the code output is correct
Debugging step through	Use the debugger to step through major portions of the code to ensure that intermediate calculations are correct
Tests with range of inputs	Test the model with an appropriate range of inputs to ensure the expected outputs are generated
Check order of calculations	When working with extreme values, check the order of calculations to minimize the truncation error
Compare results to other models	Compare the results to those in the literature or to simple models to get a sense of the expected outcomes
Check output values	For each test run, make sure the resulting outputs are in the expected range
Empirical model limits	For empirical model, ensure that the forecasts are limited to the range of the input data

make multiple runs and examine the final output to ensure that they are producing reasonable and expected values. It may not be possible in all cases to fully know what the expected range of output values may be. However, there should always be some prior work or data that inspired the creation of the model for which some reference values are available.

By following these guidelines, you should be able to verify that your code correctly implements your conceptual model. You should also be able to explicitly define the circumstances the model can be applied and have some initial ideas of the accuracy of the model that can be tested when validating the model.

13.3.3 Validation Guidelines

Model validation is often a much more difficult challenge. For deterministic models where there are reliable experimental results, the validation may be relatively straightforward. The experimental results can be thought of as the *real world* and compared against the model results quantitatively. However, even in that case, we can only show that the model is *right* within the range and controlled conditions of the experiment. It could well be that the behavior of the system will change in an unexpected way under conditions not tested in experiments.

For models dealing with open systems, the validation process is much more difficult. An air pollution model, for example, may take into account a wide range of atmospheric conditions to arrive at an estimate of the distribution of a particular pollutant within an urban area. Even if there is a long-term monitoring effort for that pollutant in the urban area, those monitors are at discrete locations that cannot possibly reflect the full range of environmental conditions (or even the range of conditions embedded in the model). At the same time, there is probably not complete measurement of all of the related causal factors at the same locations and same scale as those represented by the monitoring network or the model's spatial resolution. In that case, our only hope is to make a substantial number of calculations to create a database of pollutant levels that are comparable to the spatial and time scales of the air pollution model. Given those limitations, we may only be able to say that on average, the model appears to be a reasonable representation of the real world.

Models such as the air pollution example mentioned earlier are also stochastic. The historical record of weather events represented by a few airport and local weather stations is used to create a probability distribution of the likelihood of various combinations of wind speed, air mass stability, temperatures, and precipitation events that will impact the dispersion of an air pollutant. We then apply that probability distribution in our model, assuming that the future conditions will mimic those in the past. This places another constraint on what we can say about the validity of such models.

Nonetheless, it is possible to use the available information to provide some insights into the validity of most models. There are both statistical and graphical approaches to validation. In Sections 13.3.3.1 and 13.3.3.2, we provide a few examples to illustrate these approaches.

13.3.3.1 Quantitative and Statistical Validation Measures

For situations in which there are reliable experimental or observational data, the simplest validation test can be the calculation of absolute and relative error for each of the available data points. The absolute error will be in the units of the measured quantities and by itself may be difficult to interpret. The relative errors are in percentages, which may be a more useful statistic.

The relative errors by themselves will not provide an overall measure of the model accuracy. Averaging those errors is not a good approach since negative and positive errors will cancel each other out. Taking the absolute value of the errors and then calculating a mean is one approach to resolve this problem. Another statistic that has been used to provide an overall measure is called the root mean square error. This provides a single summary of the errors across all of the sample points by squaring the differences and taking the square root of the resulting average as shown in Equation 13.14:

$$\text{RMSE} = \sqrt{\frac{1}{n}\sum_{i=1}^{n}\left(y_i - \hat{y}_i\right)^2} \qquad (13.14)$$

where:
  y_i is the observed values
  $\hat{y}_i$ is the simulated values
  n is the number of observations

For stochastic models that produce multiple output values that can be summarized into either spatial or time based subsets, comparisons can be made between observed and simulated mean values for each subset. Going back to our air pollution model, we could make multiple runs to characterize the possible outcomes during periods of air temperature inversions when air pollutants are held near the ground. Those simulated averages could be compared with real data from days where similar conditions existed. In addition, the distributions around those means could be compared either by calculating additional descriptive statistics such as the standard deviation and the skewness or plotted to visualize their relationships. There are also statistical tests of significance such as the difference of means test that can be applied to those distributions to test whether the observed and simulated values are significantly different.

We can also use regression analysis to provide a test of the accuracy of our simulation. In that case, we would make the observed values our dependent variable and the simulated results our explanatory variable. The resulting coefficient of determination will then provide a measure of the amount of variance in the observed data explained by the simulation model. The closer the value is to 1.0, the better our model.

There are many other possible statistics that have been used in model validation. A complete review of all of those methods is beyond the scope of the current chapter. Hopefully, the examples here provide a solid base in understanding the process of model validation.

13.3.3.2 Graphical Methods

Quantitative measures for model validation are rarely employed without also examining graphical comparisons of observations and simulation results. There are several reasons for this. First, one or a few statistical measures may not fully represent the distribution of the errors in the model. A graph will show whether the errors are consistent across the observations or whether they are concentrated in a particular subset of those observations. Second and just as important, the graph will help to identify those areas or times where the model errors are greatest. That information can help to inform the modeling process and to identifying whether the errors are the result of the computation or algorithms, are related to the underlying assumptions of the model, and/or are because of an erroneous conceptual model that formed the basis of the modeling effort.

Several types of graphs can be used to visualize the model validity. Of course, the simple two dimensional plots we have been using are one of those options. For phenomena that are spatially distributed, a three dimensional graph may be more useful. Aside from graphing model results versus observations, one can also graph the deviations from various statistical analyses. For example, when a regression model is used, the residuals or unexplained variance can be plotted to identify whether the errors occur in some pattern. The residuals can also be used to identify exceptional cases. Those cases can be examined in more detail in an attempt to track down the causes of the deviation.

Finally, it is important to remember that the validation is being made with respect to the purpose of the model. The acceptability of the error will vary depending on that purpose. In addition, the model may still provide important insights into the behavior of the system that can inform future model improvements.

EXERCISES

1. Download and run the glitches.m or glitches.py program file from the book website. Run the code and observe what happens when the order of operations change. Describe the results. If you were

doing one of these calculations in a program and wanted the program to branch when the result of the calculation is zero, how would you compensate for these truncation errors? Add that code to the program.

2. Write a program to simulate the exponential growth model described earlier in this chapter. The initial population is 50 and the growth rate is 0.05. Create the model using the Euler method with a time increment of 1 and 100 iterations. Calculate the analytical solution to the equation in the same program. Graph both distributions and calculate the root mean squared error for the simulation.

3. Use one of the ode solvers in MATLAB or Python to generate a solution to the exponential growth problem above. Compare that result to the analytical solution and calculate the root mean squared error for that comparison.

4. You are charged with validating a model of the level of dissolved oxygen (DO) in a river. The oxygen level must remain about five parts per million (ppm) in order to maintain a healthy environment. You know that the level of DO can physically go from 0.0 to 14.6 ppm. The model simulates the impact of a sewage treatment plant discharge. As it flows downstream, the organic waste from the plant lowers the DO level as bacteria use oxygen to decompose the waste. Download the file DO_verify from the book website. The file has a series of model runs using different assumptions about the amount of organic waste being released and the efficiency of the environment in adding oxygen back to the stream. Verify the model using these data. Describe any model errors you can find and indicate what you would look for and/or add to the model to fix those problems.

REFERENCES

American Society of Mechanical Engineers. 2006. *Guide for Verification and Validation in Computational Solid Mechanics. ASME Standard V&V 10.* New York: American Society of Mechanical Engineers.

Bush, B. M. 1996. The perils of floating point. http://www.lahey.com/float.htm (accessed December 15, 2016).

Department of Defense. 2009. DoD modeling and simulation (M&S) verification, validation, and accreditation (VV&A). Department of Defense Instruction Number 5000.62. https://www.msco.mil/vva.html (accessed December 15, 2016).

Mathworks. 2016a. Solving ODEs in MATLAB. https://www.mathworks.com/videos/series/solving-odes-in-matlab-117658.html (accessed December 15, 2016).

Mathworks. 2016b. MathWorks products and services. https://www.mathworks.com/products.html?s_tid=gn_ps (accessed December 15, 2016).

Murray-Smith, D. J. 2015. *Testing and Validation of Computer Simulation Models.* Cham, Switzerland: Springer International.

Oberkampf, W. L. and C. J. Roy. 2010. *Verification and Validation in Scientific Computing.* Cambridge, Great Britain: Cambridge University Press. ProQuest ebrary. Web. 15 (accessed December 2016).

Schlesinger, S. 1979. Terminology for model credibility. *Simulation* 32 (3): 103–104. doi:10.1177/003754977903200304.

Wikipedia. 2016. List of Python Software. https://en.wikipedia.org/wiki/List_of_Python_software#Scientific_packages (accessed December 15, 2016).

CHAPTER 14

Capstone Projects

14.1 INTRODUCTION

The purpose of this chapter is to present a number of capstone projects that demonstrate both the modeling and programming skills that have been acquired throughout this book. Typically, the projects should be undertaken over a period of about four weeks while students work through the remainder of the modeling and programming content. The projects can be completed by groups of students or by individuals with the expected outcomes adjusted according to the number of students completing the work. Each projects require some research about the system being modeled, the programming of a basic model, the addition of one or more additional model components to relax some of the initial assumptions, model runs to answer one or more research questions, model verification, model validation where possible, and presentation of the results orally and in writing.

We provide an introduction to each of the example projects distributed with the publication of this book in the following sections. Additional project materials are available on the book website. Additional project may be added to the website over time. For each project we present:

1. An introduction to the underlying system and the related mathematical equations.

2. A set of questions to be addressed through the implementation and testing of the model.

3. A set of optional additions to the basic model.

4. An optional starting code outlining the major programming steps to implement the basic model.

5. A set of references that describe the system and models of that system that have been previously implemented.

6. Where possible, links to sources of data that can be used to validate the model.

The optional starting code is available to instructors so each instructor can decide if their students should start the project from scratch and require some additional guidance. Whether or not that code is presented also depends on the amount of time granted to complete the project as well as the nature of project and its relationship to the exercises undertaken as homework earlier in the course.

14.2 PROJECT GOALS

The objective of the projects is to build and test a model of one of the candidate systems and use that model to derive insights into the system behavior with respect to a set of proposed problems. Individuals or teams undertaking the projects will need to complete a number of activities:

1. Review and understand the underlying system concepts and the mathematical representations of the system that are given in the provided materials and references.

2. Create a basic starting code or complete the given starting code to complete the model as presented.

3. Verify that the code is providing the correct results within the bounds of the system being modeled.

4. Use the code to test the model sensitivity to selected parameters, to forecast the results through a range of prescribed conditions, and to assemble the appropriate tabular and graphics information to create a summary report.

5. Enhance the model by relaxing one or more of the modeling assumptions by adding additional components to the model, conducting the necessary runs.

6. Optionally add interactive components to the program that allows data input from the keyboard or a file, writes outputs to files, and/or automatically summarizes or compares multiple runs.

7. Validate the model against real data or the literature where possible.

8. Prepare a summary report and presentation describing all of the testing, verification, validation, and model outcomes.

14.3 PROJECT DESCRIPTIONS

Each of the projects is introduced in this section. The auxiliary materials are available in the Projects folder on the book website. The optional starting codes for MATLAB® and Python are available on the instructor's website.

14.3.1 Drug Dosage Model

The first project is a model of drug dosage and its absorption into the bloodstream. A concept map of the model is shown in Figure 14.1. Medicine is ingested as oral doses one or more times per day with each pill containing a particular dosage. Those doses are given at discrete times. Assuming a 16-hour day during which medicine could be taken, the doses can be assumed to be distributed evenly over that time period.

Once a dose is taken, that amount is resident in the intestines. For each increment of time, part of that dose is absorbed into the bloodstream at an adsorption rate, adds to the amount in the blood plasma, and reduces the amount in the intestines. Once absorbed, the medicine also leaves the body at each increment of time depending on the half-life of the drug and the excretion rate.

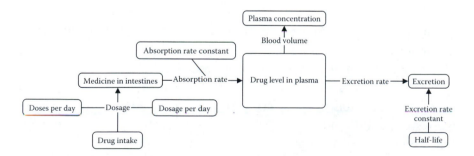

FIGURE 14.1   Concept map of drug dosage model.

At any of the time periods, the plasma concentration is a function of the plasma level divided by the blood volume of the person who took the drug. For this modeling effort, we assume that there is a concentration level goal that must be reached for the drug to have the intended benefit. In addition, too high concentration produces a toxic effect, which creates unwanted side effects. The goal of the model is to find the dosage and timing of doses to achieve the medicinal level and stay below the toxic level of concentration in the blood plasma. There are also some optional explorations and model extensions that can be completed as part of the exercise.

This simple model of drug dosage is called a one-compartment model of drug dosage. The model assumes that the drug achieves instantaneous distribution throughout the body. It also assumes that the level of the drug in the blood plasma is the same of that in the organs and tissues that the drug is used to treat.

Use the model to vary the doses and dosage per day such that the overall goal is obtained. You should document your code and provide graphic and tabular outcomes to show how you arrived at the appropriate levels and timing of the dosage. These should be contained in a written report and prepared for presentation along with any of the additional options you added to your model from the list of possible additions given in the detailed assignment description.

14.3.2 Malaria Model

Malaria is one of the most devastating diseases, and it is a leading cause of death in tropical regions of the world. Mathematical models of the disease help public health professionals to have a better understanding of disease transmission and to identify effective measures for the prevention and elimination of the disease.

Malaria is a vector-borne disease spread through the bite of a female Anopheles mosquito. Thus a model of the disease requires two submodels—one representing the life cycle of the mosquitoes that spread the disease and one representing the cycle of the disease in humans. Figure 14.2 is a diagram from the Centers for Disease Control and Prevention that illustrates both of the cycles.

An infected mosquito injects parasites called sporozoites into the bloodstream. They hide in the liver for a period of one to two weeks where they divide asexually and mature into merozoites. Those are released into the bloodstream and infect red blood cells. Some of the parasites mature into gametocytes, which can be ingested by a mosquito to pass the disease

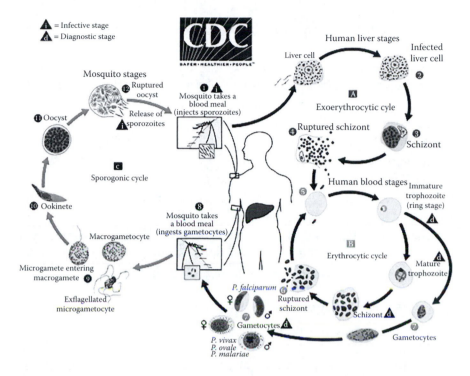

FIGURE 14.2 The malaria infection cycles. (From https://www.cdc.gov/malaria/about/biology/.)

back to that disease vector. The male and female gametocytes merge sexually to produce sporozoites that are transmitted back to humans in the salivary glands of the mosquito.

This project will implement a simplified model of the malaria disease cycle applied to a village in a region with malaria. Assuming that there is no latent period for infection, the population of humans in the village can be divided into three groups: healthy villagers, sick villagers, and immune villagers. For any given time, the change in the number of healthy villagers depends on the numbers of births, deaths, infected, and recovered villagers. Change in the number of sick villagers depends on the number of infected villagers, recovered villagers, villagers who gained immunity, and deaths of sick villagers. Change in immune villagers depends on the number of villagers who gained immunity and the number of deaths in immune villagers.

The life cycle of the malaria parasite in mosquitos is little simpler since infected mosquitos end with death, with no recovery or immunity.

For any time period, changes in the number of healthy mosquitoes depend on the numbers of births, deaths, and infected mosquitoes, and change in the number of infected mosquitoes depends on infected and death of infected mosquitoes during that period of time. Review the references for this project to provide additional background on the equations used in modeling malaria.

The detailed assignment provides a list of starting equations and parameters for you to use in creating this model. You should start by reviewing the references provided for a more detailed review of the life cycle for malaria and the nature of efforts aimed at controlling its spread.

You should create the basic model using those instructions and then prepare a report describing the nature of the findings for the basic model and for any of the optional model additions you implement.

14.3.3 Population Dynamics Model

One of the intensively studied populations of predators and prey is the wolf/moose populations of Isle Royale in Lake Superior. As this is an island, the relationships between the populations are not nearly as complex as in other predator/prey populations. As an example, you can compare the conceptual relationships of the species on Isle Royale versus Yellowstone National Park as shown in Figure 14.3.

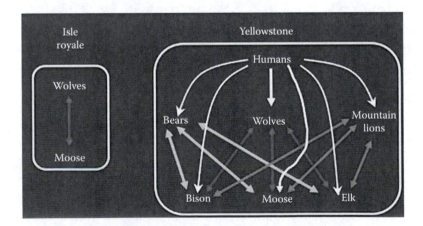

FIGURE 14.3 Comparison of predator prey with Yellowstone National Park. (From http://www.isleroyalewolf.org/overview/overview/the_setting.html, 2014.)

The classic predator–prey model is based on the Lotka–Volterra equation for population change (Wikipedia, 2016):

$$\frac{dx}{dt} = x(\alpha - \beta y) \tag{14.1}$$

$$\frac{dy}{dt} = -y(\gamma - \delta x) \tag{14.2}$$

where:
x is the number of prey
y is the number of predators
The equations represent the growth rate of the populations over time
t represents time
α, β, γ, and δ are constants representing the interactions of the population

Conceptually, this model can be visualized by the concept map shown in Figure 14.4. Each of the populations is governed by its own birth and death rates. In addition, the availability of the food supply of moose impacts the birth rate of the wolves, whereas the number of wolves impacts the death rate of moose.

Your project will be to build a simple model of the wolf and moose population starting with detailed instructions and parameters given on the project website. Once you have built this model, you should see the cyclic nature of the population growth and decline associated with this system. Test the sensitivity of the model to changes in the moose birth and death rates. For the given birth rate, what does the death rate

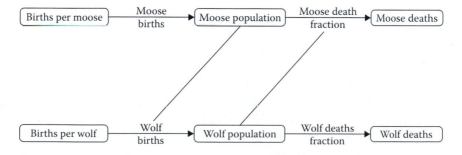

FIGURE 14.4   Concept map of Moose–Wolf population dynamics.

need to be to provide a more stable population of both species over time (e.g., a range closer to 250 rather than 600–800). How does this change if the moose birth rate increases by 50%? Explore the sensitivity of the wolf birth and death rates in the same way. Add one of more of the optional exercises based on the research results of the Isle Royale research project. Use these simulations to prepare a summary report and presentation.

14.3.4 Skydiver Project

This project focuses on modeling the fall of a skydiver and plotting their position and velocity in free fall and after they open their parachute. Modeling the motion and location of a person or parcel exiting an airplane or helicopter with a parachute must account for a number of physical laws and related environmental conditions. If these are not taken into account properly, a skydiver might open his or her parachute too late to slow down before reaching the ground, a person seeking to reach a particular ground location could end up far away from the target position, or a parcel dropped to reach particular recipients on the ground would be lost. The challenge of this project is to create a model that accounts for several of the variables that affect the flight of a skydiver or parcel dropped from an airplane—predicting their velocity and location.

Two sets of forces act on the skydiver. First, there is the acceleration due to gravity moving the skydiver toward the ground. Second, there is the resistance of the air acting in the opposite direction. The basic equations we used in the ball model previously can be used to calculate the acceleration. Remember that acceleration is defined as the change in velocity with respect to time. Applying Newton's second law of motion:

$$F = ma \qquad (14.3)$$

where:
  F is the force
  m is the mass of a body
  a is the acceleration

Stated differently, acceleration is directly proportional to force and inversely proportional to mass. In the case of the skydiver, the acceleration is the acceleration due to gravity. This is approximately −9.81 m/s². It is negative when upward direction is considered to be the positive direction. The total force in the downward direction is then the mass × acceleration.

This accounts for the downward force but does not consider the drag associated with the friction of the air. The drag will be a force in the opposite direction, which is related to the density of the medium the object is moving through. To account for this, we need to add the drag equation:

$$R = 0.5\,DPAv^2 \tag{14.4}$$

where:
    D is the drag coefficient
    P is the air density
    A is the cross-sectional area of the object
    v is the velocity

The vertical position of the skydiver is then calculated as:

$$y = v_0 t + \frac{gt^2}{2} \tag{14.5}$$

where:
    y is the vertical position
    v_0 is the initial velocity
    t is the time
    g is the acceleration due to gravity (9.81 m/s²)

However, the drag is not constant over time. There will be some air resistance of the skydiver based on how much they spread their limbs to catch the air. That force will continue until the ripcord is pulled and the parachute deployed. At that point, the parachute will introduce additional drag to slow down the descent.

The initial parachute model should account for these two forces assuming a constant air density. The model should then be used to experiment with the time and location of free fall with the deployment of the parachute to determine whether the skydiver will land safely. There are then other components that can be added to remove some of the model assumptions. The details for those optional simulations are provided in the downloads for this project along with a list of references that can be used to choose an appropriate set of parameter values for the simulation.

14.3.5 Sewage Project

The health of a stream can be measured in many different ways. One of the most critical conditions needed to maintain a healthy aquatic community is the level of dissolved oxygen in the stream. The dissolved oxygen (DO) is the source of oxygen for all aquatic plants and animals. Natural levels in a stream can go from 0 to 14.6 parts per million. In water, this measure by volume is also equivalent to the measure by weight—milligrams per liter, because of the molecular weight of water.

When sewage is released into a stream, the organic material is decomposed by bacteria. The bacteria consume oxygen, thus depleting the amount of oxygen available to the other biota in the stream. To minimize those impacts, municipal sewage treatment plants use one of several processes to greatly reduce the amount of organic waste left in the effluent before it is released into a stream or lake. Nevertheless, the remaining waste causes a reduction in the dissolved oxygen level.

The simplest model of dissolved oxygen (DO) was developed in the 1930s by two civil engineers named Streeter and Phelps. This Streeter–Phelps model depends on the relationship between oxygen demanding wastes, measured by biochemical oxygen demand (BOD) and the rates of deoxygenation caused by its decomposition. This rate is balanced against the rate of reaeration related to adding oxygen back from the atmosphere. Those rates, in turn, are related to the physical conditions in the stream and the stream temperature.

Right after the waste is discharged, the deoxygenation rate exceeds the reaeration rate causing a decline in the DO level. This occurs until the first set of waste is decomposed. Then the stream begins to recover as the reaeration rates exceeds the deoxygenation rate. The process produces an oxygen sag curve looking something like Figure 14.5. This figure shows the oxygen level as a function of time as bacteria decompose the waste load creating an oxygen deficit while the atmosphere reaerates the stream.

The Streeter–Phelps model for this process is actually a differential equation:

$$\frac{dD_t}{dt} = k_1 L - k_2 D_{t-1} \tag{14.6}$$

where:
D is the dissolved oxygen deficit over time
L is the concentration of organic matter requiring decomposition
k_1 is the coefficient of deoxygenation
k_2 is the coefficient of reaeration

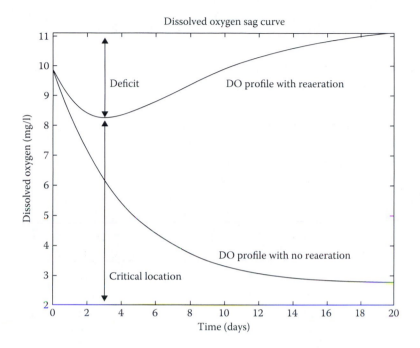

FIGURE 14.5   Components of the oxygen sag curve.

Thinking about this equation with respect to each increment of time, it says that the oxygen deficit for the current time period is first, dependent on the deoxygenation coefficient (a constant for each section of the stream) times the volume of waste remaining to be decomposed. The deficit is off-set by the reaeration rate times the oxygen left after the last time increment. That is because the greater the difference between the concentration of oxygen in the stream and the concentration in the air, the greater the rate of diffusion of oxygen back into the stream. The constant for reaeration also varies by stream segment as each segment may have a different amount of turbulence and different streambed and therefore more or less mixing with the air.

This modeling project will have you build a DO model using a modified Streeter–Phelps formulation and using that model to evaluate some policies for changing a sewage treatment facility discharging into an example Ohio stream. You will test the sensitivity of the model to changes in the parameters and then undertake one or more optional modeling tasks to examine the impacts of an increased waste load on the stream and validate the model against some sample data from the same stream.

14.3.6 Empirical Model of Heart Disease Risk Factors

For this project, you will build an empirical model of the risks for heart disease and use that model to simulate the impacts of changes in the price for cigarettes and other behavioral changes on the occurrence of the disease. The project starts with the use of a 1999 dataset from the Centers for Disease Control that shows the major risks for heart disease and the rate of disease for each state in the United States. You will use this to build a multiple regression model where the response variable is heart disease and the possible predictor variables are the risk factors. The risk factors include smoking, obesity, lack of physical activity, and the positive impact of eating fruits and vegetables.

Once you have a model where you have identified all of the statistically significant predictors, you will use that statistical equation to build a model that allows the examination of the impacts of policies such as increasing cigarette costs on smoking rates and then on heart disease. Additional optional analyses include assembling a similar dataset for 2014–2015 to see if any progress has been made over time in reducing risks and heart disease rates.

14.3.7 Stochastic Model of Traffic

This project will involve the collection of local data that help to build and test a stochastic model of the travel time to work that was completed in Chapter 2. In that exercise, you may recall that there were two distinct paths from home to work that were simulated—one using only local streets and one using the highway. Each street and highway section travel time was a linear function of average travel speed, distance, and traffic control devices that predicted the travel time for that segment. The time added by traffic control devices (stop signs and traffic lights) was input as constant parameters for each segment where they appeared.

In reality, the amount of time it takes to traverse an intersection with a traffic control device depends on the random occurrence of conflicts with other vehicles and status of traffic lights when a traveler arrives at an intersection. In this exercise, you will gather data for a few representative intersections in your own community and use the data you gathered to create a stochastic representation of what could happen at each intersection. This assumes that the hypothetical trip in the exercise is in your community and that the sample data you collect apply equally to all similar intersections in the simulated trip.

For intersections that have four-way stop signs, the amount of time it will take to clear the intersection is related to the number of cars that arrive at the intersection at the same time as well as their order. For our purposes, we will not consider the order of arrival directly but will instead collect information on the number that arrives at our sample intersections at the same time and the time it takes for the last car to clear the intersection as well as the frequency of those events during a representative hour.

Traffic lights have a similar impact with the addition of the timing of the green light period for the target direction of travel. The total time at such an intersection will be related to the cycle time for the traffic light and the number of cars stacked up in the direction of travel at a red light. When the light turns green, the more cars that were stacked at the intersection, the longer it will take for the last car in line to clear that intersection. Again, you will pick one or two sample intersections in your community and compile data on the light timing and time to clear the last car over a representative hour.

Once you have the relevant data, you will need to create two functions that use a random number generator to pick the time to clear the intersections at traffic lights and stop signs respectively. You can then run the model a large number of times and then analyze the range of times it may take for that simple trip. The instructions for this project provide some guidance on taking and analyzing a sample and integrating the resulting distributions into your stochastic model.

14.3.8 Other Project Options

You may have your own ideas for a project with similar scope. If so, you should create a proposal for undertaking that project that includes a short description of the model purpose, an initial conceptual model, and the initial mathematical representation you will code based on the available literature. That should then be discussed with your instructor before you proceed with the project.

REFERENCE

Wikipedia, 2016. Lotka-Volterra Equations. https://en.wikipedia.org/wiki/Lotka%E2%80%93Volterra_equations. As viewed on December 15, 2016.

Index

Note: Page numbers followed by f and t refer to figures and tables, respectively.